ALISA BRYCE is a soil scientist with a BSc in agriculture from the University of Sydney and a Masters from the University of Cambridge. She likes digging holes, analysing soil, playing with soil, writing about soil and patting dogs.
alisabryce.com.au

Grounded

How soil shapes the games we play, the lives we make and the graves we lie in

Alisa Bryce

TEXT PUBLISHING MELBOURNE AUSTRALIA

The Text Publishing Company acknowledges the Traditional Owners of the country on which we work, the Wurundjeri people of the Kulin Nation, and pays respect to their Elders past and present.

textpublishing.com.au

The Text Publishing Company
Wurundjeri Country, Level 6, Royal Bank Chambers, 287 Collins Street, Melbourne Victoria 3000 Australia

Published by The Text Publishing Company, 2022

Cover design by W. H. Chong
Page design by Rachel Aitken
Typeset in Sabon LT Std 12/18.5pt by J&M Typesetting
Map by Simon Barnard
Index by Geraldine Suter

Printed and bound in Australia by Griffin Press, part of Ovato, an accredited ISO/NZS 14001:2004 Environmental Management System printer.

ISBN: 9781922458650 (paperback)
ISBN: 9781922459947 (ebook)

A catalogue record for this book is available from the National Library of Australia.

For Mum and Dad

Contents

By way of introduction

IF I'M AT a party and I don't want to talk to someone I've just met, I'll tell them I'm a soil scientist.

'Sorry?' Their brow furrows.

'Soil,' I say, pointing to the floor. 'Earth, dirt...' I don't like using the word dirt but it is sometimes necessary.

'Ah,' they nod. 'Soil.' Some say it twice, as if trying it out for the first time.

'A soil scientist,' I repeat.

They may or may not ask what a soil scientist does, before the inevitable point where silence falls and they excuse themselves to go get a drink.

If I do feel like talking, I'll say I'm a writer, though the outcome is usually the same. They ask what sort of writer, I say science writer, they ask what sort of science, I say soil.

Then they go get a drink.

It's understandable. Soil is an unusual topic. The probability of randomly meeting someone who can converse on the subject is small. There are only about a thousand soil scientists in Australia, so the chances that you have met a soil scientist are even smaller. I didn't know it was even a career option until my second year at university, when studying a Bachelor of Science in Agriculture.

Soil is, of course, usually examined through the lens of agriculture (and if not agriculture then climate science), which is also understandable. Over ninety-five per cent of our food comes from the soil. Most people who work with soil work in agriculture, and most books about soil are in some way about agriculture—different cultivation methods, environmental impacts, historical developments and so on.

This book is different. It does not discuss the merits of various farming practices, and there is no advice on growing tastier tomatoes. Nor does it deal with carbon sequestration, or any of the other topics that were popular at the time of writing. This is not a book about climate change, erosion or environmental degradation. There are plenty of other books available on these topics. Instead, this book explores the other ways the sticky, macabre and quite frankly fascinating world of soil underpins much of life, culture and society.

Follow the mysteries in the Crime chapter and pour yourself a robust glass of shiraz to read the Wine chapter. Slap on a face mask as you read about clay and beauty. Travel back in time with me, 3.7 billion years, to a world where soil and life were just coming into being. Imagine what it might have been like for the soldiers deployed to conduct covert soil sampling on the Normandy beaches in January 1944. Ponder the concept of a nutrient afterlife rather than a soul afterlife. And okay, let's talk about farming—but how it might be done on Mars.

Researching this book has been a lot of fun (though I do wonder if I am now on an ASIO watchlist, and if so, I would like to assure the good people of the various security forces that my questions about the blast radius of sixteen kilos of TNT and the use of lime to dissolve human bodies were purely academic). But a book that covers many topics can, of course, only offer a glimpse into each, and for every chapter I included, I had to leave at least one out.

There isn't the scope to delve into the nuances of what is a very complex science, and this is obviously not a textbook. It is a taster, a teaser: a fun introduction to something you probably didn't realise was so fabulous. The wonderful world of soil.

Life

IMAGINE YOU ARE lost in a vast wilderness. No phone reception, no satellite phone: no way to call for help unless you are skilled with smoke signals. You have tried shouting, crying and praying to gods you didn't really believe in until now. The only response is the occasional bird call. You are completely alone. Someone may find you eventually, but until then the only way you are going to survive is by using the resources at hand. (It helps if you have imagined yourself into a forest-type wilderness rather than a desert.)

First, the basics. You will need to find water, food and shelter. If you are lucky, you'll find running water, but do you have anything to store it in? You can forage, if you know which plants are edible and which are poisonous; hunting means building a trap or maybe a spear using sticks; for fishing you'll need a line, a hook, a net—what will you make them from? You will need a fire: for warmth, to cook what you have caught and to fend off whatever's doing all that howling (probably just the wind...). Fire needs kindling and a spark. If you are really good at this survival business, you might fashion a mat out of reeds and start carving a bowl or spoon, because luckily you had a knife in your pocket when you got lost.

It will be tough. But everything you need to survive is somewhere in your surroundings and almost all those things are there because of the soil. The branches for kindling, the shrub that bore the (thankfully non-toxic) berries, and the reeds for the basket.

Humans lived this way, off local resources, for most of human history. It is only very recently that you could just pop out to the shops and buy things you don't need, made on soil in countries far, far away.

Life can only exist because of the soil. That sticky, gritty stuff beneath your feet is the foundation of life on Earth. Except for water and oxygen, everything we need we get from the soil—and even then the soil filters the water. If the soil suddenly vanished, land plants would quickly die off, and with them almost everything else. But don't worry, you wouldn't suffocate from lack of oxygen.* You'd die from starvation first.

---

* While plants do produce oxygen, scientists estimate (and their estimates vary quite a bit) that fifty to eighty per cent of atmospheric oxygen comes from phytoplankton in the ocean.

What would we be without soil? Certainly not a planet worthy of the name Earth.

Soil is fascinating and full of paradoxes. It is one of Earth's oldest resources, yet soil science is one of the youngest sciences. The soil is both biotic and abiotic, both alive and dead. It is 'dirty' but also an expensive beauty treatment. We are utterly dependent on the soil, yet we take great pains to banish it from our lives.

I have noticed in recent years, however, that soil is slowly making its way into general conversation. Not very often, and not in much depth, but it is there. Every year on 5 December, World Soil Day, there's a smattering of articles about the importance of soil. Gardening has had somewhat of a resurgence, particularly during the pandemic. There are even one or two popular science books about soil now gracing bookshop shelves.

If I asked you 'Do you know what soil is?', I imagine you would say yes. Everyone knows what soil is. But take a few minutes to think about it. What exactly is this stuff that delights and pains us? Not even soil scientists can agree on an exact definition. Like love, like home, soil is hard to define.* It means different things and has different uses to different people.

It's where plants grow, of course, but what of the soil that became paintings, or pots or building foundations, or tunnel walls, or a mud mask—even a meal?

---

* Thanks to Greg Retallack for this wonderful description.

Soil is usually described as the top layer of the Earth, the living skin stretched across this rock of a planet. This is largely true, but what about soils that were once at the surface but are now covered by layers of new soil blown in by a storm or exploded out of a volcano?

It is often described as dark or brown, but that is only some soils, or only a small part of a soil. Cave paintings would be very bland if all soil was brown.

Is there a difference between soil and dirt? At what point does soil become dirt? Or is it just a matter of nomenclature? Maybe dirt is to soil as weeds are to plants: the right thing in the wrong place. Just as a tomato plant in a wheat paddock is a weed, soil on the carpet is dirt. And what of potting mix? It grows plants, yes, but is it a soil?

The more you think about the soil the more complicated it becomes, which is partly why it is so fascinating.

The best description I have found so far is one proposed by Brian Needelman in a *Nature Education* article. His description encompasses the thousands of types of soils across the globe without trying to specify location, depth, colour or use.

Brian says soil is a recipe with five ingredients: minerals, living things, organic matter, air and water.

The mineral part—from coarse sand to tiny silt and clay particles—comes from rock, weathered over time into smaller and smaller pieces. Living things are the billions of microbes, insects, plant roots and everything else that has set up home in and on the ground. Organic matter is mostly the remains of once-living things—leaves, roots, microbes, corpses. Most organic matter comes from plants, the leaves and twigs that drop and decompose, and from the exudates that roots release: sweet sticky stuff for soil microbes to enjoy.

As far as I can tell, the only non-debatable ingredient is the minerals. If a soil dries out completely it is still considered soil. If you compact a soil and squash all the pores out of it so there is no space to hold air or water, is it still soil? I think it is. Depending on who you ask, there may be soil on other planets—even in the absence of organic matter or life forms.

But in this book I am going to consider soil as having at least minerals, organic matter and life. Because on Earth, at any rate, life can only exist because of the soil, and soil as we know it only exists because of life.

The very long shared history of soil and life

They go back a long way together. Longer than soil and plants; even longer than soil and oxygen. The relationship between soil and life began billions of years ago, when the two began a process of coevolution that led in time to the huge range of soils and life forms on Earth today.

Once upon a time, long before time was a concept, the Earth was a mass of elements swirling about the universe. Eventually the heavier elements sank to the centre of this cosmic tornado, colliding and fusing to make the Earth's core, which is mostly iron and nickel. Around the core formed the mantle, a nearly three-thousand-kilometre-thick layer made mostly of rocks rich in magnesium and iron. Intense heat melted some of the rocks, creating liquids that rose to the surface then cooled enough to crystallise and form a first attempt at a crust. About four billion years ago, the Earth began to cool and the lighter elements such as oxygen, silicon and aluminium on the outer rim of the molten Earth solidified into another crust. It's still not entirely solid: even today Earth's crust is not one single piece but a series of huge plates resting above the boiling mantle, occasionally bumping into each other and causing an earthquake or forming a mountain.

Then at some point—no one knows exactly how or when, though it's at least 3.7 billion years ago—the Earth started to get soil. Freezing and thawing, wetting and drying, grinding and cracking: the rocky crust slowly eroded into rock dust. This early soil was very different from the soil we know today. It was an inert, gravelly dust, like something you'd expect on the Moon or Mars. It was granular and to the untrained eye probably looked like soil, but it struggled to hold water and could not support larger life forms. There was nothing by way of organic matter—either in it or above.

Then, as early life forms got to work, the rock dust evolved too.

The idea that life came from soil is quite common in religious traditions—the Judaeo-Christian origin story has Adam created from dust, for example, while the Inca god Viracocha made people from clay, and the Ancient Egyptian god Khnum created children from clay—but has fallen out of favour in science.

The more common theory is that life started in water, an idea that has gained great popularity since the 1950s when

Miller and Urey made organic molecules from inorganic components by sparking water and gases with electricity.* One (greatly simplified) version of early life and soil formation proposes that very early life forms lived at the water's edge in great big microbial mats, photosynthesising, respiring and dying in hordes to leave organic matter (themselves) to mix with the weathered rock particles on the shore.

This 'soil' had not weathered enough to acquire the sticky clays that help clump soil into aggregates with pores and channels in between. It did not have the structure to hold on to water or organic matter, which were probably washed straight back out again. But in time, through enough weathering and organic matter accumulation, life crept onto land.

Palaeontologist Greg Retallack argues that life began in the soil, supporting Scottish chemist Alexander Graham Cairns-Smith's idea that clays were the precursor for life. In this theory, early chemical weathering of the Earth's rock crust left behind sticky clays that were not carried away by wind or water—and life evolved in the nooks and crannies of these sticky clays. Clay is arguably a better place than the sea for life to take hold as it offers better conditions for complex organic molecules to form. They like variety, and clays come in a variety of shapes with a multitude of compartments for molecules to form in as they have wet-dry cycles,

---

* Miller and Urey simulated the Precambrian atmosphere, putting carbon dioxide, ammonia, methane and hydrogen in a closed system with some water, then sparking it with electricity to simulate lightning. This made organic molecules (amino acid and sugars) in the water at the bottom of the flask. They didn't create life, but did make some components of it.

and essential trace elements such as iron and molybdenum. The sea, by comparison, is a dilute solution that is relatively uniform. So organic life evolved in the clay, eventually separated itself from the clay and learned to live in and on the soil. Molecules appeared that could use the sun's energy and nutrient elements, and so clay became a source of food—and the two began their long, long history.

It's a contested theory, and maybe we will never know the answer—just as we might never know exactly what life form appeared when. But we can imagine how soil grew and blanketed the earth once life had been established and got busy growing.

Early non-vascular land plants like mosses and liverworts colonised damp rocks, slowly eroding them and releasing clays and nutrients. Fungi wove across the land, also mining rocks and decomposing detritus. Scientists believe mycorrhizal fungi were critical for vascular* land plants to develop, helping them survive before they had developed enough of a root system to make it on their own. When plants with more advanced roots developed, they pushed into the soil, creating pores and channels for water and air to flow. Fungi and plant roots still have this relationship today: the fungi offer improved nutrient uptake, protection from pathogens and resistance to stresses such as drought, while the plants release

---

* Vascular plants have conducting tissue (xylem to move water and nutrients up through the plant and phloem to transport sugars around the plant) and are the more familiar plant types with roots, shoots and leaves. Having vascular tissues allows plants to grow taller than non-vascular plants like moss.

exudates from their roots to feed them.

Lichens probably played a decent role. Lichens are very hardy creatures, and today are just about the most adaptable life forms on Earth, able to survive on bare rock by excreting organic acids that erode the rock and release nutrients.

When insects and other soil fauna appeared, they burrowed and churned, mixing the organic matter in with the minerals, further helping the soil develop a structure more suitable for land plants. Plant roots continued to find their way through existing soil into the solid rocks below, seeking tiny fissures and cracking the rock. In time, the world stopped being a barren and probably rather boring place and became an incredibly varied and rich world, full of billions of types of life and a vast array of different soils.

One particularly interesting soil–life relationship, a young one by soil standards, occurs in the beautiful grassland soils— sometimes called black soils, prairie soils or Mollisols—most prevalent in North America, South America and Eurasia. These deep, organic, nutrient-rich and incredibly fertile soils are some of the most productive in the world. They form under the dense tangled root systems of prairie grasses, which pump copious amounts of carbon into the ground. Thirty million years ago, Mollisols did not exist. Just as Earth's first soils remained rudimentary until life appeared, these soils could not develop until a bigger life form appeared: grazing animals. As grazers evolved hard hooves and teeth designed to tear at the grass, the grasses developed rhizomes, underground stems, nodes and other adaptations that helped them cope with the assault. By about eight million years ago, these soils and grasslands had expanded into the humid regions of North America, Africa and Asia, and coincided with global cooling.

Soils are incredibly variable. There may be a soil that is brown all the way down from the surface until you reach rock somewhere underground, but it is not the norm. Soils come in all colours: red, black, brown, yellow, green, grey and even blue. Some are shallow, some are deep. Some are stripy or mottled or impressive masses of walled colour. Soils range from coarse sand to heavy clay and everything in between. There are rocks, roots and different-shaped aggregates, and all these parameters can change the deeper you dig. One soil scientist I know describes soils like a sandwich with different layers of thickness, colour, consistency, texture and structure—ham, cheese, tomato and mayonnaise. Another says that looking only at the topsoil, about the top fifteen centimetres, is like only looking at the head of a bird. You're ignoring the fact that it's got wings, legs, a tail, a body and feathers in a range of colours.

Even the standard soil-profile diagram is a necessary oversimplification, showing only five layers (or 'horizons', in soil science). There's an organic (O) horizon at the top, then topsoil (A horizon), subsoil (B horizon), parent material (C horizon: the rock from which the soil came) and bedrock (D or R horizon). In fact, soils exhibit an incalculable variation in number of layers, layer depth, colours, minerals, structure and texture (the proportion of sand, clay and silt). These differences stem from the factors that affect how soils form.

Soils vary so much across the globe because of all the different climates, topographies, rocks and life forms. Over time, climate, topography and living things work on all the different rocks in infinite variations to create the ever-changing soils underfoot. They are still forming and changing today. Expose the same rock to different climates and you get different soil. Soil is usually shallower at the top of a sleep slope and thicker at the bottom. As plants grow, their dropped leaves and roots are added to the soil. Soil forms faster in warm, damp climates where there is more microbial activity. Soils with iron turn bright red when well oxygenated, green if they are underwater.

Organisms play a huge role in how fast soil forms and what properties it will have. The types of plants that grow can affect whether a soil becomes more acidic or alkaline, and whether a soil will have more or less organic matter. Soil properties such as pH and salinity in turn affect what plants can grow. The type of microorganisms, such as bacteria and fungi, affect how quickly nutrients cycle and what plant materials are broken down. Soil critters like worms and insects burrow through and churn the soil, making pores and channels. Humans can, and these days often do, have the biggest impact of all.

—

Hawai'i—soil formation in action

Soil as we know it could not form without life, and life could not have evolved without soil. Today the soil supports all life above it, and harbours one-quarter of all biodiversity on Earth (at least so we believe: no one really knows how many types of microbes are in the soil so it's hard to quantify).

Back in the early days of the Earth, making soil took millions of years. But as life forms appeared and climates developed, the process got faster and faster.

Today the soil–life relationship develops much more quickly because higher life forms already exist. Seeds and spores can easily blow or wash in and take hold, while more amenable climates help them establish faster.

These days soil can form in decades if the conditions are right. But a few general rules still apply. You can't grow higher plants until you develop soil. Put another way, you can't grow big trees on bare rock. The odd plant might take hold, but you won't find a forest on an area yet to develop soil.

The best thing to do for some essential research into modern soil formation (global pandemics and other catastrophes permitting) is to hop on a flight to Hawai'i. As a chain of islands, each younger than the last, it is a brilliant place to see the progression from bare rock to forest in action. Hawai'i sits above a huge tectonic plate—the Pacific Plate—that is slowly moving the islands north-west. Under the chain is a volcanic hotspot that causes eruptions, spewing new lava and making the islands grow. It's currently located beneath the Big Island, which is growing acres of volcanic land each year. As the plate moves north-west the hot spot spews lava and creates a new island, while the older islands weather, to the point that some of the islands at the top of the chain are now little more than sandbars. So on four- to five-million-year-old Kauai you can see old, developed forests; on the Big Island there are swathes of bare new land recently formed by volcanic eruptions and lava flows where life has not yet had a chance to take hold.

One example of converting a bare lava flow to a rainforest in Hawai'i goes something like this. Once the lava has cooled, blue-green algae and some early mosses settle in lava fissures, taking hold in the crevices where moisture can pool and there is some protection from the heat as the sun warms the bare rock. After about four years a nitrogen-fixing lichen called *Stereocaulon vulcani*, sometimes called snow lichen, appears: a grey, fuzzy thing that makes the landscape look like there has been a dusting of snow. The mosses, lichen and algae work away at the rock, excreting the acids that dissolve the rock, releasing essential nutrients including iron, and helping the rock become soil. As nitrogen-fixers, both lichen and algae can capture nitrogen from the air and make it available in the soil for themselves and other plants to use. So in time, the fissures where the lichen live start to fill with rock dust and organic matter.

Pioneer shrubs like *Dubautia scabra* and 'ōhelo 'ai (*Vaccinium reticulatum*) can use this dusting of soil in the lava cracks and bunker down in what is still effectively just rock. With its pale green leaves and light red berries, 'ōhelo 'ai looks pretty but is very tough, able to withstand up to twenty-five centimetres of ash falling on it—just in case there is another eruption.

After fifty to a hundred years, young stands of *Metrosideros* trees move in. These trees can grow to twenty-five metres tall but, in this undeveloped soil, they will only become small shrubs. Because the trees have short, stubby branches, there is still plenty of light to reach the ground, which is ideal for a mat-forming fern, uluhe, to move into the open bedrock. By covering the ground, uluhe stops other tree seeds from germinating. With the increasingly diverse range of plants growing in the slowly transforming lava rock, dropped leaves and fronds start to accumulate, the beginning of an organic topsoil.

As the trees grow, they lose their stubby branches and form a canopy above. The uluhe fern, preferring light, retreats from the newly shaded environment and a more shade-tolerant, taller tree fern (*Cibotium*) takes over. The organic layer in the soil gets deeper as insects, plant roots and all the wonderful soil life work together to make a soil that suits them. Leave these friends alone for a few hundred to a thousand years, and you will get a rainforest overlying a Histosol: a soil with at least forty centimetres of organic materials (more than twelve per cent organic carbon) in the top eighty centimetres.

On different rock in a different climate with different plant species, different soils would form. In any case, the soil and life forms above it eventually reach a sort of equilibrium. In very rainy areas the minerals in the soil might leach deeper underground, leaving a bleached layer (like the soil at Stalag Luft III mentioned in the War chapter). In higher rainfall areas in Hawai'i, the rain leaches calcium and potassium deeper into the soil, leaving an acidic rootzone that the forests can tolerate and adapt to, but which needs management if the land is cleared for farming. In cold, wet areas, you might get peat. In warm, wet rainforest areas, life moves so fast that plants and microbes quickly churn through any nutrients dropped from the forest. Test these soils and they will seem hungry, with most of the nutrients sucked up into the lush forest vegetation above.

Where there is soil, there is life

Most of Earth is fairly hospitable to life and there are only a few regions where it cannot take hold. Among the most inhospitable places are the dry valleys of Antarctica: a mere smattering of life, and a surface that tends to spark debate on whether or not it is really a soil.

New Zealand soil scientist Megan Balks has made nineteen trips to the cold south to study Antarctic soil. 'When I got my first study trip,' she says, 'I thought *what soil*?' Did Antarctica even have soil, or was it just rock dust? But a little case of scientific incredulity wasn't going to stop her going on an all-expenses-paid adventure. 'When someone offers you a trip to Antarctica, you go.'

Megan's mission was to assess the impact of human life on Antarctica by seeing what evidence lay in the soil. Upon arrival at Marble Point, about eighty kilometres from the

McMurdo Dry Valleys, all she saw was rocks. No plants, no animals, nothing except grey rocks with occasional patches of salt. This was the desert pavement, a layer of rocks packed together at the surface, left behind after all the smaller particles have been blown away or settled deeper between the rocks.

'We started off looking at oil spots and fuel spills and understanding what happened to it. We were digging up old oil spills, some dated back to the 1950s. You could dig a hole and find liquid oil spills sitting on top of the permafrost at about fifty centimetres deep in the soil,' says Megan. That first trip was in the 1990s, which means the liquid oil had been sitting there for over forty years. 'You could soak a piece of paper in the liquid then set it alight,' she says.

What Megan also found were the very limits of what we could reasonably call soil. Below the rocky surface was a coarse, gravelly sand with almost no organic matter. 'So by some people's definition this is not a soil, it is a sandy gravel.'

We could ask the same question of beach sand. Is it soil? Or is it just sand?

At the time, some definitions of soil insisted upon the presence of organic matter (which generally only accumulates if the soil is supporting plant growth). Here there were no higher plants, nothing to drop leaves or twigs or other sources of organic material that in a more favourable environment would eventually decompose and build up an organic layer in the soil—nutrients for microbes and plants that would support more plant growth above. Yet even in this medium,

which barely qualified as a soil, there was life. Bunkered
down among the rocks were moss, lichens, even some spring-
tails and tiny red mites, barely visible without a magnifying
glass. Green tinges on the underside of marble rocks hinted
at algae and at endolithic cyanobacteria (endoliths live inside
rock, shells and other hard substrates). And there were likely
other tiny life forms that found a niche in cracks and crevices
where there is barely a taste of water and light—just enough
to survive.

Researchers from the University of Colorado recently collected samples of what appear to be sterile soils from Antarctica.
Microbes can live in the most inhospitable environments, yet
here there seems to be...nothing. (The researchers admit,
though, that they can't definitively say the soil is sterile; just
that if there *are* organisms in the soil, they haven't managed
to convince them to grow.) This 'soil' was found just over
2100 metres high, on top of a mountain poking out of hundreds
of metres of ice, and full of nitrates and toxic perchlorates. I
don't know if they tested for organic carbon. It raises the same
question—were the samples soil or just crushed rock? Without
life, without organic carbon, is it even soil?

Perhaps what they found is the true boundary between
soil and life.

Antarctic soil, though often very old, is still in the very early
stages of formation. Antarctica is a world where the climate—
the cold, the dark, the aridity—has slowed down the fifth
soil-forming factor: time. With life forms in such short supply,

soil formation largely depends on physical and chemical processes—repeated freezing and thawing, wetting and drying, rocks grinding against each other—to slowly break the rock down into smaller and smaller particles. Some minerals react with acids, oxygen, carbon and water to cause chemical processes that also wear away at the rock. But the climate, an important part of soil formation, keeps life, and consequently biological weathering and soil formation, at bay. As we've seen, soils develop much faster in areas where it is wetter and warmer, where moss, algae, lichen and microbes can get to work turning the gravelly sand into something more recognisable as soil.

One exception in Antarctica is where the penguins live. Four of Megan Balks' nineteen trips to Antarctica were to Cape Hallett and Cape Royds, where there are Adélie penguin colonies. Soil sampling among curious penguins was a little more exciting (and far more adorable) than in the dry valleys.

'Little penguin chicks are very cute and fluffy,' says Megan, 'but they lack the water-repellent adult feathers to keep them warm and dry. If they get wet, they don't survive the freeze.' So the dutiful parent penguins build their nests on a pile of stones. Stone size is very important: big enough to be useful, small enough for a penguin to carry in its beak. About two to five centimetres is optimal.

'They build their nest, and they poop everywhere so the nest becomes a stuck-together mess of stones and penguin poop and eggshells,' says Megan.

The penguins come back to the same place every year, where the nest is often completely buried in penguin shit, and they need to get more stones to build it up again.

'Over time they build a whole soil profile which we call penguin soils, or ornithogenic soils if you want to be more technical. Penguin soils consist of the nest of stones, penguin guano, the remains of birds that have died and the eggshells.'

There is more than enough organic matter in these soils to keep the soil-classification people happy. And all that guano changes the soil nutrition and microbial communities. The poop is a source of nutrients such as carbon, nitrogen, phosphorus and silicon, and harbours poop-dwelling microbes, and the extra organic matter means the soil can hold more moisture from the summer snow melt. These soil changes mean that the microbes change, with more of the types of microbes you expect in a stomach—Bacteroidetes, Firmicutes and Thermomicrobia—than the Actinobacteria, Chloroflexi and Nitrospirae you would find in non-poop soils. If this part of Antarctica wasn't so darned cold and dry, this highly organic poopy soil would allow different plants to flourish above. As it is, it's just penguins.

Not only is penguin soil smelly to study—the team weren't very popular back at the base camp after spending their days digging around in a thousand years of accumulated shit—it also requires unusual care.

'To work in these places, you have to have environmental permits in triplicate. So we carefully put out groundsheets, excavated the profile in the right order so we could put the soil back and restored the surface so you couldn't tell where we'd been.'

As Megan and team worked, they dug about one metre deep, placing the excavated rocky soil in a careful pile beside the hole.* But penguins, while unconcerned with environmental permits, are very much concerned with collecting stones. 'The penguins spend a lot of time stealing stones off their neighbours and guarding their own,' Megan says. With thousands of penguins looking for stones every year, the beaches were pretty well sifted through for good nest-building stones. And now here, next to these apparently harmless hole-digging creatures, were piles of new ones.

The penguins were quite happy to toddle around Megan and crew, sifting through their spoil pile for useful stones. This did not cause any issues and made the day rather more enjoyable. The problems arose when the crew went away for lunch and came back to find some of the feistier penguins had claimed the entire soil pile and were 'sitting on top of it like the king and queen of the castle'.

Although these penguins were mighty pleased, they couldn't keep the new nest. Environmental permits and all that. The team had to shoo them away and return the stones to their place beneath the surface. 'There were,' reports Megan, 'some very disappointed penguins.'

---

* The point of this research was to understand patterns of penguin habitation and how it influences soil development, and to see if there was any evidence of human activities.

Where death and life meet

When we die, our bodies become the grass, and the antelope eat the grass. And so we are all connected in the great circle of life.

   —Mufasa, in *The Lion King*

There is a set number of elements on Earth. About twenty of them, including nitrogen, phosphorus, calcium and sulfur, are part of every single living thing: everything from a carrot to a whale to a virus—and you. In this busy world full of different life forms, life just keeps turning over.

The partnership entered between soil and life forms billions of years ago persists today. The soil is a portal between the living and the dead, recycling the elements to turn one living being into the next. And the critters that live on and among the soil particles, in the cracks, pores and channels, are the masters behind it all. They constitute the soil ecosystem, and it is glorious. The jungles we see above ground are deserts compared to those underground. Earthworms sail through the topsoil, leaving nutrient-rich casts in their wake. Protozoa and nematodes fend off pests (or become them), and graze on herds of bacteria and fungi. Fungi enter agreements with plant roots, offering protection in return for food. Though we cannot see most of these creatures with the naked eye, what they lack in size they make up for in numbers. The saying goes that there are more microbes in a single teaspoon of soil than there are people on the planet.

When a leaf falls to the ground, or a lion exhales its

last breath, the soil ecosystem begins its work. Insects and microbes begin a feeding frenzy, seeking out the nutrients the recently deceased no longer has use for—carbon, nitrogen, phosphorus and more. Microbes on and in the body begin to digest their once-living host. Soft tissues break down into liquids, salts and gases. Heterotroph bacteria get to work, breaking down the eyeballs and liver into their individual elemental components. Aerobic bacteria do this via respiration; anaerobic bacteria use fermentation, producing gas such as methane and ammonia in the process. Some bacteria break down proteins, releasing ammonium in the process. Others convert the ammonium to nitrite, then nitrate, the major form of nitrogen used by plants. Insects attract other insects, colonising the once-living thing in waves. The noble earthworm helps out by incorporating organic matter into the soil, making it easier for microbes to do their work. All creatures of the underground play a part. In time the once-living thing is marked by a cadaver decomposition island (CDI), a nutrient-rich area of soil where plants might grow better and soil life is more abundant. (In time. At the start, the CDI can cause fluctuations in soil chemistry, inhibiting plant growth and leaving bare patches on the ground.)

The living thing is now well past the realm of the biotic. It has become the soil, ready to enter the world of the living, but in a different form. When viewed from a soil perspective, decomposition is quite beautiful. Not an end, but a transition. It is the idea behind natural burials, which we'll hear more about later: the idea of a nutrient afterlife rather than a soul afterlife. Since the beginning of humans about 108 billion people have lived, along with so many other living things that, were it not for the soil, the world would be covered in layers and layers of corpses. Sometimes decomposition is fast, sometimes it is slow, but give it enough time and once-living things eventually become soil.

Compost is the perfect illustration of death becoming life. A carrot goes into the compost heap and in a few months is completely unrecognisable. Put the compost back on the soil and it provides nutrients for things living in the ground and above it. Rain falls; water fills the pores, mobilising elements into a plant-friendly nutrient drink with nitrogen, phosphorus, potassium and other tasty elements. The plant roots start drinking and the elements make their way into the grass above, the nitrogen and iron pumping dark green hues into blades of grass that will now grow twice as fast. You cut the grass, the blades fall, and biological reincarnation continues.

The elements that make up everything on Earth are in a constant state of flux. They cycle through the soil, plants, animals, water and the atmosphere, changing form and location but never being created or destroyed.

Living things battle for these elements, persisting for a few days or decades, then giving them back to the Earth. This is the way it has been since the beginning of life. Every living being is reincarnated...or at least recycled. Your eyes contain atoms that were once, perhaps, a dinosaur's toenail, your eyelashes maybe once a blade of grass. The carbon in your body could have been buried deep within the Earth, then shot to the atmosphere by a volcanic eruption, floated around for a decade or so in the atmosphere before being dissolved in the ocean, then taken up by a fish that was caught and eaten by your mother.

Enjoy the time, space and elements you rent on this planet. After about eighty years, if you're lucky, you'll give them back for something or someone else to use. Perhaps you'll become a vase or a piece of art; a tree or a flower. Maybe you'll be carried around by air currents or drift across oceans, bumping into other particulate beings on your travels.

Everyone leaves a legacy in something or someone else.

It is the soil that allows us to do so.

Making soil 1: fun and profit

LIFE FORMS MAKE soil. Humans are the only life form to make it consciously, deliberately and often at eye-watering expense. This is most strongly evident in the worlds of high-end sport and property development where soil, rather than being formed slowly *in situ* over hundreds or thousands of years, is manufactured to a formula in a matter of weeks.

The soil that supports facilities like sports fields and public gardens often falls into one of two camps: highly engineered, or whatever happened to be around at the time. Passive recreation areas like open parkland, local sports fields and roadside gardens are usually planted in whatever soil happens to be there, or whatever the developer dumped there, which will probably be a mix of topsoil and subsoil excavated from elsewhere during development.

But as soon as you try to grow plants in some way they don't normally choose to grow—on a roof, for example, or as a perfectly flat field for athletes to kick or hit a ball on—natural soil doesn't cut it. Beneath these plantings lies engineered soil, designed to keep the plants alive under extraordinarily difficult conditions. Urban soil scientists are rare, but theirs is some of the most interesting work. They get to start from the beginning, designing a new soil from scratch, by asking: what do I need this soil to do and how can I make it do it?

You can 'make' soil, or something to grow plants in, by combining the usual five ingredients—minerals, organic matter, life forms, air and water. Getting the mineral and organic-matter parts right is the most important. When they're right, the air and living things (such as microbes) can move in of their own accord, and adding water is easy enough.

I used to work for a company that regularly made growing media. We would take various ingredients such as drill mud, sand, sugarcane waste and various composts—big baking scoops full of them—and blend them in different ratios. We would try out various mixtures, send the samples to the lab for testing, then tweak the recipe based on the results, trying to balance the physics and chemistry. Permeability too low? Maybe a little less drill mud; more compost if we needed more organics, or a different type of compost if one was too rich in certain nutrients. We almost always managed to come up with a formula suitable for some kind of growing medium, whether it was a vegetable-garden blend or a native-garden soil mix.

Once the recipe is worked out, the big machines of the commercial soil manufacturers blend and tumble various ingredients—sands, composts, maybe some loam, diatomaceous earth, perlite (there's a lot to choose from), pushing them around with front-end loaders, mixing with excavator buckets—until there is a reasonably homogenous mix. The soil is then loaded into the back of a truck, driven to the site, unloaded and spread. In the case of green roofs, which need special lightweight media to grow plants on rooftops, the 'soil' is sometimes sucked up with a long hose and spat out or blown onto the roof.

There is a great example of manufactured soil at Barangaroo in Sydney. Until 2008 the site, at Millers Point on Sydney Harbour, was a working port. Once the port was moved out of the harbour and further south, the old port was locked off and remained an empty and unsightly strip along the fore-shore. The state government decided to redevelop the concrete mass into a residential and commercial space, incorporating about six hectares of parkland complete with walking trails and lookouts over the harbour.

The landscape designers wanted to hark back as far as pos-sible to what the headland might have looked like in the late 1700s, studded with sandstone blocks and graced with the trees and shrubs native to the shoreline: scribbly gum forests on the hilltops, blackbutts and blue gums on some mid and lower slopes. By the time a detailed plan had been devised, the site would use eighty-four different plant species that were indigenous to the Sydney region at the time of European settlement.

Much of the parkland is built on land that, fifteen years ago, was a sheet of concrete extending all the way to the water. Now if you dig beneath the happily growing trees you will find soil: a very coarse yellow soil manufactured specifically for the project and currently sustaining over 75,000 plants and just over two acres of grass. Much of it was made using crushed sandstone spoil excavated during the development. The cost of trucking soil off-site and disposing of it in landfill is at least fifteen dollars per tonne (plus the EPA landfill levy), and more than two hundred dollars if the soil is contaminated.* Then you have to buy in new soil for the plants—at least sixty dollars per cubic metre for anything reasonable. Barangaroo used about 50,000 tonnes (about 32,000 cubic metres) of site-excavated soil. At a conservative estimate of fifty dollars per tonne to go to landfill and fifty dollars per cubic metre to buy new soil, that's just over four million dollars saved. Fancy saving a few hundred thousand or a million bucks? Reuse the soil on site where you can.

However, simply finishing the development, spreading around whatever soil is left sitting in piles and hoping plants

---

* Be cautious about signs advertising 'clean' soil. Unless the seller has a test sheet with full contamination testing, you don't know what is meant by 'clean'.

will grow in it is not wise, although it is quite common. Digging up and pushing soil around destroys the soil structure, and then you get a compacted medium that plants usually struggle to grow in. The death toll for new plants in many new developments is routinely twenty per cent and up to fifty per cent if things really go wrong. And when you're spending six figures or more on plants, a fifty per cent survival rate isn't good enough. Simon Leake, who designed the soil for Barangaroo, tells the story of a Sydney local council that spent tens of thousands on plants for an indigenous species garden display, only to have most of them die. When they replaced them, a whole lot died again. If they'd spent a thousand dollars up-front and checked the soil, they would have known it was far too alkaline or 'limey' for the plants they had chosen, and that they wouldn't survive. They would also have avoided the embarrassment of wasting all that ratepayer money, well over a hundred thousand dollars.

For the Barangaroo project, Simon's team set about designing a soil mix that they would be confident the plants could grow in. This meant looking at which plants the landscape architects had chosen and working out what sort of soil they liked.

Did they prefer acid, alkaline or more neutral pH? Could they tolerate salt? Were they hungry species who needed rich soil, or did they prefer a low-nutrient environment? Phosphorus or not? There's a common belief that native Australian plants don't like phosphorus: true for some species, but not all. Many, such as *Acacia floribunda* (a type of wattle) and *Melaleuca radula* (graceful honey-myrtle), grow happily in phosphorus-rich soil.

Unfortunately, there is no catalogue available detailing exactly how much of every nutrient every plant species can tolerate. So the team set up trials. They grew what they thought were the most sensitive species, most likely to curl up their toes if the soil wasn't right—plants such as *Banksia marginata* and *Banksia spinulosa*—in various combinations of crushed sandstone, sand and compost, the basic raw materials available. Balancing the physics and chemistry was a challenge. Too much sandstone clogged up the mix, making a soil prone to waterlogging. The compost was nutrient rich, and some species were unhappy to be surrounded by so much food. Then the pH was too high, making phosphorus more readily available and inducing iron deficiency.

After multiple iterations the team worked out that a mix of 47.5 per cent crushed sandstone, 47.5 per cent washed sand and only 5 per cent compost (for context, 20 per cent compost is pretty normal in many manufactured soils) allowed the most sensitive species to grow while still providing enough nutrients for the other plant species to establish. They were going to add iron sulfate to acidify the mix, making it easier for the plants to take up iron and manganese and harder to take up phosphorus, but it didn't prove necessary and for some species was dangerous! Scribbly gum took up too much manganese as the soil became more acid. Keeping the starting soil pH at about 7.5 and banking on it dropping with time as the plants took up nutrients and acidified the soil (as they do) helped balance long-term nutrient availability—not too much phosphorus, not too much manganese, and enough iron to keep everyone happy. Now, around a decade down the track, the pH is around 6 to 6.5 and still dropping. There are only one or two nutrient-deficient plants in the land-scape, and both of them are Proteaceae, a family well known for this.

That example may make it sound unlikely, but designed and manufactured soils are quite common. Potting mixes are made this way. As green roofs and green walls become more popular, so has the demand for lightweight substrates that can support plant growth without compromising the integrity of the structure beneath. You can't just dig up soil from the ground and spread it over a roof. As soon as it rains the roof is likely to collapse, or at least crack and leak.

Next time you're walking down the street, pay attention to the street trees. Unless they are very old, they have almost certainly not been planted directly into the soil. Trees are not stupid, and they *are* very determined and strong-rooted. Their hunger for oxygen, which is usually more abundant near the surface, is why they often crack the pavement (or break pipes that carry oxygen, water and nutrients). Newer street trees are usually planted in a pit specially designed to let the roots grow happily alongside all the services that run underground with them. It might be structural soil which is made of eighty per cent gap-graded crushed stones and twenty per cent soil, the stones graded in such a way that they won't pack together and are strong enough to support the weight from pavement and even cars while giving the tree a chance of surviving with food, water and air. Or there might be a specialist cell system, designed with the same aim in mind but usually made out of plastic, with dedicated channels for services to run through and with a lower chance of slippage.

There is an entire industry dedicated to manufacturing the soils or growing media for all these situations. I was on the Australian Standard 4419 review committee; the standard sets out the criteria for what properties manufactured soil should have. We spent years debating the minimum standards for the physics and chemistry for garden soils, landscape soils, wetland soils and soil under parkland turf.

Where we didn't dabble was the high-end stuff: the elite sports fields, the golf greens and cricket pitches. These industries have dedicated specialists and unique soil constructions of their own.

Elite fields and golf greens

If you peek under the turf at the Melbourne Cricket Ground (or almost any elite sports field) you would be forgiven for thinking you had dug a hole near the beach. The turf is growing in sand. But I'll wager you've never seen grass of this lush green quality growing at the beach, and for good reason: this is not in any way a 'natural' set-up.

Groundskeepers, particularly those responsible for elite sports fields, have the occult task of convincing grass to grow where it otherwise wouldn't. Sand does a very poor job of holding water and nutrients. A soil with more clay, for example a loam, can hold more of both, meaning the grass can go longer without food and water top-ups. Even keeping grass alive, let alone healthy and thick, on a bed of sand requires a great deal of maintenance. The grass is constantly hungry and thirsty, and needing regular fertilisation and irrigation. It's also subject to frequent mowing and foot traffic, both of which stress the turf so it needs even more food and water to recover.

The higher the expectations of a sports field—more hours of play, absence of bumps or lumps, perfect-looking turf for the TV cameras—the sandier the soil beneath, to the point that elite sports fields are ninety to ninety-five per cent sand. No A-grade sports-field manager would hold up a handful of rich black soil and call it 'good': its use would spell the end of their career.

And not just any sand: the sand beneath elite fields has been carefully chosen to include a certain range of particle sizes and shapes. No more than three per cent clay, and ten per cent total silt plus clay. Any more than that and the pores (gaps between the particles) can get clogged, slowing water infiltration so that players have to wait longer after rain before they can get back in the field. The sand is also poorly graded, meaning most of the sand particles are one particular size. Well-graded soil, which is what engineers usually want, has an even distribution of different-sized particles that pack together neatly. The smaller particles fit inside the gaps between the bigger particles, making a firm surface that is great to build on or cap dams and stop the water from seeping away. On a sports field the particles should pack together tightly enough to make a firm surface, but not so tightly as to cause unsightly depressions or undulations that might trip a star player.*

As for the shape of the particles, that isn't too often an issue. But it was for an elite field in Queensland back in about 2009, when they had problems with a very slippery surface. A quick check of the sand under the microscope showed the particles were very round, like little ball bearings that slid easily over each other, making the surface very unstable. If the particles are too sharp or angular, on the other hand, you can get distorted grass roots, and slower drainage because they can pack together more tightly than rounder ones.

The idea behind all the sand engineering is to keep water flowing through the soil, to reduce waterlogging, compaction and downtime. Sand particles are much bigger than clay, able to stack together in a way that leaves more gaps (pores) between the particles for water and air. The bigger pores drain quickly; the smaller pores hold water for a touch longer. A soggy ground deforms under foot traffic. Athletes trip, golfers can't putt straight, complaints ensue. Plus there's all the work to get the surface flat again.

The flip side of such rapid drainage is that the rootzone

---

* Ideally, you want sixty per cent of the particles to be between 0.25 and 1 mm in diameter, less than three per cent of them should be bigger than 3.4 mm, less than ten per cent should be smaller than 0.15 mm and there should be a spread of particle sizes across other ranges. This should create a rootzone that balances drainage, water retention and firmness.

dries out fast. Thirsty turf not only looks bad on TV but it wears out and the surface becomes less stable, slippery even. Athletes fall over, complaints ensue.

Luckily, scientists have worked out how to keep some water in the sand by creating a perched watertable. High-end fields often have at least two layers: the sandy rootzone at the surface overlying a layer of gravel. Water 'perches' in the sand above the gravel.

To understand how, we need to take a look at the physics of water draining through soil. It's fascinating and a little counter-intuitive, as two opposing forces are at work:

1. Gravity, which pulls water downwards through the soil, and
2. Capillary action, which pulls water upwards through the soil.

When all the soil pores are filled with water, the soil is saturated. As the soil drains, gravity moves water out of the largest pores first, leaving water in smaller pores. Cohesion and capillary action keep the water in the smaller pores. The soil reaches an equilibrium, known as 'field capacity', when the force of gravity is about equal to the cohesive and adhesive forces keeping the water in the pores.

Because a gravelly soil has big pores its cohesive and adhesive forces are weaker than the sandy soil with smaller pores, so the water stays in the sandy soil rather than draining into the gravel.

When the soil is saturated the weight of the water pushes

it down and the water can drain into the gravel, preventing waterlogging. This is why putting gravel at the bottom of a pot 'for drainage' won't actually help with drainage unless the potting mix is saturated.

It's an elegant and relatively efficient design. Elite sports fields still have decent irrigation schedules, but with engineering and physics working hand in hand they don't need to be complete guzzlers.

Constant haircuts and regular trampling create hungry turf that needs more food and water to recover. Groundskeepers have an array of techniques for keeping turf alive, but at the end of the day it is a living thing that can only handle so much abuse. Even with the best sandy rootzone, irrigation and nutrition program, at some point the grass simply cannot recover from the constant wear and tear, no matter what the groundskeeper does. You might have seen this in your local park, perhaps on a much-walked path, where turf is constantly re-laid but before it can establish it is trampled to death again. The only options then are to cut back the hours of use or turn to artificial turf.

I once did a site inspection at a very expensive boys' school where the groundskeeper was in trouble for not being able to keep the grass looking good. He was facing an impossible task: the field was subject to about sixty hours of foot traffic a week, including cadet training with cadets stomping their boots into the ground and grinding the turf to pieces. He was never going to be able to keep that grass growing, and he knew it. I was brought in to be the impartial third party to write a report saying what he had been telling his bosses all along: either cut down the use or get artificial turf.

Cricket pitches

At the opposite end of the spectrum to golf greens and elite sports fields are cricket pitches, designed to be less like beach sand and more like concrete. What makes pitches so interesting—apart from the media storm they can cause if they're not up to scratch—is that the groundskeeper has to make conditions as difficult as possible for grass to grow, then convince grass to grow.

Pitches around the globe have reputations for suiting different types of bowling: Australia and South Africa for the quicks, the UK for seam and swing bowlers, and the Asian subcontinent for spin. A multitude of factors affect these differences, but they can be broadly categorised into climate, the type of soil used on the pitch, and how the pitch is managed. It is the amount of clay, the type of clay, the level of moisture in the soil, the compaction level, and the species and coverage of the turf that affect pace, bounce and spin so profoundly.

Higher-end pitches have three main layers. The exact materials and depths vary for each pitch but are generally a drainage layer at the bottom, one hundred to two hundred millimetres of sand above, then two hundred millimetres of clay at the surface. This top layer—the clay—is what takes all the work. Clay is the surface layer of choice because you can compact it to make a very hard surface; if it's solid enough for building foundations it's firm enough to bounce a ball.

The amount and type of clay in the pitch affect how hard it will set. More clay means a harder pitch. Australian pitches are about sixty to eighty per cent clay and tend to contain smectites—clay minerals that swell when wet and shrink when they dry out. They dry out harder than clay minerals that don't swell. The high clay content and shrinking clay means we can get a very hard, compacted pitch that's fast with good bounce.

Wickets on the subcontinent have more sand: the wickets are less cohesive because sand is not sticky like clay. After a few days of play these wickets turn into dustbowls, busting open and sending up clouds of dust when the ball hits them. Sand is more abrasive than clay, which helps grip the ball so it spins.

UK wickets are about thirty to forty per cent clay and are usually made from kaolinitic clay minerals that don't shrink–swell. Therefore the pitches don't dry out as hard, making them slower. The upside is that these lighter pitches drain more quickly, so you can get back on after rain faster than you can on a high-clay Australian pitch.

Pitch management

The crux of managing a pitch is balancing water and compaction to get the surface hard and dry. Groundskeepers aim to compact the soil to the point that water will only filter in at a rate of about three millimetres per hour. (Compare this to golf greens, where they want water going through at a rate of at least 150 mm/hr.) As for moisture, the pitch needs to be dry, but not too dry. Too dry and the pitch crumbles and breaks, a dangerous situation in that if the ball hits the cracks it can bounce up erratically, at risk to the batter or the close fielders. In 2015 a match between Australia and New Zealand at Blacktown International Sportspark was called off by lunch on day two. The pitch was already dry and cracking before the match began, and after only a short amount of play the New Zealand coach described it as 'a jigsaw with half the pieces missing'. The pitch was too dangerous for fast bowlers so they opted for spin bowlers instead. Jarrod Bird, an ex-cricketer who has spent twenty years preparing wickets across the globe, chalks it up to a very heavy workload of AFL leading up to the game, as well as bad weather in the week before. Both teams agreed the pitch was unsafe for play.

At least five days before a match starts, groundskeepers begin a cycle of irrigation and rolling to compact the pitch. Irrigation aims to get water evenly throughout the clay—no easy feat as the soil is already compacted, so it takes an hour for only a few millimetres to infiltrate. Regular, short watering is necessary. Too much at once and it just runs off the sides and the pitch dries out.

Once it's been wet up, the pitch is rolled, usually about one to two days after watering when the surface is 'tacky'. Every curator will have their own way of judging when the pitch is ready for rolling. Curators often only need to push their knuckle into the pitch (or walk barefoot on it) to know what needs to happen to get the surface ready for play. If the groundskeeper can make a slight indentation they're good to roll. Experience comes into play, because naturally the size of your thumb and the pressure you use will affect how much of an indentation you can make.

Rolling at the right moisture content is critical. Too wet and you can damage the pitch as the soil sticks to the roller and can make a wavy surface. It also won't compact well because water doesn't compress. Rolling compacts the soil by squishing the pores, and that won't work if all the pores in the soil are filled with water. A too-dry pitch also won't compact, as you can imagine: when you walk on dry clay you don't leave a footprint, whereas on wet clay you might sink down to your ankle.

Then the cycle continues, irrigating and rolling over many days, making sure the right amount of water goes evenly into the pitch, and using the correct roller weight, speed and direction, until the surface is deemed good enough for play. For a one-day match this is as hard and dry as possible. How hard? It varies. Some literature says 1.6 grams per cubic centimetre. Seasoned groundskeepers, using their knuckle, know what feels right.

For a five-day Test match, it's different. There needs to be some moisture in the clay and some green grass starting out, since you can't irrigate during a game. Providing it doesn't rain as the days wear on, the pitch dries out to become faster and more bouncy, and the ball seams around less, adding some variety to the game. By days two and three the pitch is generally supposed to be ideal for batting. By day four it's getting dusty and starting to crack as the clay dries and shrinks, which slows the pitch down: it's less bouncy and better for spin bowlers.

As groundskeepers cycle through the irrigation and rolling they are also trying to keep the turf alive and covering the pitch evenly. This is no easy feat with the turf now growing in very inhospitable conditions. Rolling stresses the grass by a) physically flattening the blades into the soil and b) removing most of the soil pores the roots use to access water and air. A general rule in soil science is: if there is less than ten per cent pore space you're not growing happy plants. Wickets usually go below this, down to five per cent. One set of guidelines from the UK suggest that the more compacted the wicket, the more uneven the root growth. Roots tended to grow nearer the surface, which would limit the plant's ability to extract water at depth. Seems fair enough: if I was a grass I'm not sure I'd be bothered trying to put roots down into what is effectively concrete. On top of that, the turf is mowed as short as possible, leaving little leaf blade to photosynthesise and grow.

Life is hard for the turf, and it takes a lot of effort to keep it alive. Why have turf at all, then? Why not just play on the clay?

Well, grass impacts play in multiple ways and that makes the game more interesting. First, it keeps the pitch conditioned: as grass grows it sucks up water through its roots, drying the soil around it. Uneven grass means uneven drying, some hard and some softer patches, a non-uniform pitch and unhappy players. With no turf at all the pitch would dry unevenly as well. Grass roots act as a sort of binder, strengthening the pitch and reducing cracking. And the grass is something of a cushion that stops indents forming in the surface.

Second, groundskeepers work with the qualities of the grass itself: they may try to roll the leaf into the wicket, which creates lubrication between the ball and the wicket, which increases the pace. Thick grass changes the momentum of the ball, acting as a cushion that slows down the ball and takes movement out of the delivery. The angle at which the seam of the ball hits the grass changes how the ball pops up. More grass makes a surface that is more damp, and that imparts more lateral movement to the ball. Broadly speaking, grassier pitches favour bowlers because they can move the ball around more; less grass favours batters. That's why, as a game wears on and the grass degrades, the pitch moves from favouring the bowlers to working more for the batting team (at least before the pitch starts to break up and the bowlers get on top again).

Watering and rolling are a big part of pitch management but far from the only factors. Climate, season, foot traffic— they all come into play. In Australia, pitches quicken up over summer, when turf is growing well and the soil and weather are warmer; they tend to be slower over winter, when the weather is cooler. On fields that host football matches as well as cricket, the trampling degrades the pitch and makes more work to help it recover before the next cricket match. Some stadiums have turned to drop-in pitches to stop the winter ball sports and overexcited concertgoers messing with the pitch.

Get the pitch wrong, especially on an elite field, and the social ramifications are dire. The curator at Sabina Park in Jamaica was threatened with dismissal and hounded by the media and the public; he even contemplated suicide after preparing what has been dubbed the 'worst pitch of all time'. The Sabina Park Test match of 1998 between England and the West Indies lasted only 10.1 overs and was the first match in 121 years of Test cricket to be called off because of the pitch. The third ball nearly hit the batsman (Mike Atherton) in the face and the rest of the short game was a minefield, with batters hopping about to avoid injury but still getting hit. When a fast bowler can deliver the ball at 150 kilometres per hour you don't want that very hard ball heading towards your face—or any part of your body for that matter. According to one account the West Indies bowlers tried to bowl consistently to avoid harm to the English batsmen, but the pitch was too unpredictable.

This was more than twenty years ago. Imagine how bad it would be today with all the armchair groundskeepers on social media, experienced in giving opinions but not in pitch preparation, baying for your blood.

With cricket wickets underlying careers and the entertainment of millions, clay sources for wickets are unsurprisingly kept secret, and are very expensive. Sources are limited—the wicket clay around Sydney called 'Bulli' soil has been all used up now. As always, when something in demand becomes more scarce the price goes up.

'It's like gold,' Jarrod Bird tells me. 'Suppliers charge an absolute fortune.' Which may sound like profiteering, but is perhaps a good example of a soil that is properly valued. Specific soils are highly coveted and consistency is crucial. When repairing a wicket, more soil is pressed into the footmarks and divots. This new soil needs to be as close to the wicket soil as possible in terms of particle size and minerals. If it isn't it will behave differently—it might shrink or swell more, crack differently, or dry faster or slower, making unpredictable patches across the surface.

The worst thing, according to Jarrod, is delamination, sometimes called a linear break, where the clay cracks horizontally between the two different clays, or clay layers are separated by dead organic matter. This can happen during wicket repair if there are grass clippings between the old and new soil. The repair soil won't sit properly, being unstable and a weak point in the pitch. If the ball hits these weak patches it can rear up in all directions. Curators try to stop this happening by flooding the wicket to make the clippings float, then brushing the wicket clean before topdressing the wicket with new clay.

Sport plays a huge role in Australian life. Our sportsmen* are paid absolute fortunes to run, jump, kick, bowl and putt on

---

* Not the women, unfortunately.

precisely engineered soils and very high-maintenance grass. Yet how much thought is given to those who keep that grass in pristine condition? The most famous groundsman on the Australian cricket scene is Nathan Lyon—and he's famous for being a Test bowler, not a (former) groundsman. Given that a talented curator has the power to swing a game if they want to, I sometimes wonder if they ever get caught up in betting scandals. Jarrod tells me that yes, in some countries the bookies do approach the wicket managers hoping they'll do something to tilt the game in favour of one team over another. He also assures me this does not happen in Australia.

At least not that he knows of.

Crime

THE DISAPPEARANCE OF Ying Holding and Chung Mi Ng in 2000 was a turning point in Rob Fitzpatrick's career. That September, Rob came home from his son's soccer game to find an urgent message on the answering machine. The police were looking for two missing women (a mother and grand-mother). They had a suspect (the son, Matthew Holding), and a car boot full of muddy and bloody objects: a shovel, boots, jade bracelet, knife, pine post, towels and bedding. The shovel suggested Matthew had been digging. But where? The police caught up with him near Moonta, two hundred kilo-metres north-east of Oakbank in South Australia, where the family lived. The women could have been anywhere along that two-hundred-kilometre stretch, either injured but still alive, or dead.

With their suspect refusing to talk, the police needed help
narrowing down their search area and turned to the evidence they had. Michael Heath, the crime scene investigator,
thought the soil on the shovel could be from local roadworks.
The site operator said no, it wasn't the same type of soil. But
it had come from somewhere—and who better to ask about
mud than a soil scientist?

Rob wasn't a forensic soil scientist then; it wasn't a career
option at the time. But he dutifully pulled out his microscope
and Munsell soil colour book—the soil scientist's version
of a Pantone colour chart but with mostly browns, reds
and yellows—and his team at CSIRO got to work. To the
untrained eye a smear of mud on a shovel might not say much
except 'wash me'. But Rob's team managed to narrow down
the search area to a quarry in the Adelaide Hills.

A reasonable amount of soil was stuck to both the
back and front of the shovel, while the edges were clean,
meaning it had been used in wet soil. The soil compacted
in the handle housing on the back of the blade ominously
suggested it had been used to flatten soil: you've finished
digging, you turn the shovel over and whack at your pile of
earth. The soil was yellowish pink, it looked like a mixture
of iron oxides and clay, and was acidic (low pH) but not
salty. The lack of plant roots, leaves or other organic bits
you'd expect to find in the topsoil suggested it wasn't from
the surface. Quartz grains in the sample were angular, not
smooth. You'd expect surface grains to have smoother edges
from being weathered and moved around, so that added

weight to the theory the soil was from deeper in the ground.

Putting all these clues together, Rob and team started narrowing the search area. They needed to find a wet area without topsoil, with yellowish-pink soil that was acid but not salty. Police had been searching the Yorke Peninsula, where Matthew was found, but the soils there are alkaline (high pH) and often saline.

Using soil maps and his own experience—Rob had conveniently been part of ongoing soil-research programs in the area—Rob directed police to a low-lying section of Oakbank quarry, an industrial gravel quarry in an acidic and low-salt region in the Adelaide Hills. This working quarry, the topsoil stripped away long ago, was only a few kilometres from the suspect's family home. Rob collected soil samples next to a large pool of water and compared them to the samples on the shovel, boots and bracelet.

Same colour; acid, not salty; similar mineral composition. The soil on the shovel probably came from the quarry so the police searched it—again. (This was the second search. Before Rob was involved they'd found tyre tracks there similar to those from Matthew's vehicle, but that first search yielded nothing further.)

The second looked fruitless, too. Time was marching on, rain making the search treacherous and cumbersome, the investigators limited to digging by hand as excavators became mired in the sticky soil. As the days turned into weeks and no bodies or other evidence were found, it appeared Rob was wrong. After all, Oakbank was an operating quarry—would

Matthew have risked being discovered while hiding the bodies?

But one particularly dogged police officer returned daily to see if the local foxes had uncovered anything in the night. His hunch paid off: on 9 October, nearly three weeks after the women went missing, the officer saw a hand sticking out of a low-lying wet patch in the quarry, just fifteen metres from the large pool of water Rob had pinpointed as the search starting point. A second body was uncovered the next day, fifty metres away. Matthew Holding was convicted of murdering his mother and grandmother and sentenced to life in prison.

This case founded forensic soil science in Australia, re-establishing a science that had fallen out of favour when the sexier DNA analysis came along. Since then the team at the Centre for Australian Forensic Soil Science (CAFSS) in Adelaide have worked on over two hundred cases, including murders, counter-terrorism investigations and recovery of stolen goods. Their reputation is such that the police have elicited five confessions just by threatening soil testing. In one case, a man had murdered a partner and buried the guns, and was staying very tight-lipped. The police needed help from the CAFSS, but first they told the defence they were going to get their 'soil forensics guy', and it was just a matter of time before they found the guns. The man confessed.

Questionable alibis

'The forensic scientists got ripped apart in court,' Rob says, his voice still with a hint of South African accent after thirty years in Australia. Even though I am sitting within arm's reach, he speaks as if I'm on the other side of the room. Perhaps it's the effect of two decades giving evidence in court: a tendency to project. Rob is explaining why forensic soil science is a relatively new career option. After all, soil is a wonderful form of evidence. It is tiny and sticky and it doesn't decay. So why wasn't it commonly used in court until more recently?

Until the late 1980s, soil evidence was simply one part of a general forensic investigation. Then there was a rapid rise in the demand for technical expertise on every subject—fingerprints, toxicology, podiatry, DNA, languages, you name it. If you wanted a forensic opinion, you got an expert. Defence lawyers could easily discredit the general soil knowledge offered by a forensic scientist simply by pointing out that it wasn't backed by any specialist credentials.

Thankfully, any soil that was collected during investigations in these dark ages of forensic soil science can still be used and analysed now. Unlike cloth, paper or DNA, most soil samples do not degrade once collected, especially if stored carefully in dry conditions, and soil in the landscape does not change very fast. In one cold case from 1988, the body of a drowned teenage girl had been found in a wetland. The police had their suspect but could not link him to the crime scene. The CAFSS used samples of the grey and yellow clayey soil stuck to the suspect's trackpants that had been held in evidence for seventeen years. By carefully comparing them to samples collected from the wetland where the body was found, they conclusively tied the suspect to the crime scene.

Clay, in particular, is very sneaky. Each clay particle is less than 0.002 millimetres in size and invisible to the naked eye. (To imagine clay particle size, scale everything up. One grain of clay is about the size of a baby mouse, a grain of silt is like a beagle and a grain of sand is a large African elephant.) The clay particles are so tiny that even the most cautious criminal would probably overlook the few grains Rob and team need to make an analysis.

It is also very sticky, adhering to your shoes, clothes and fingernails. Wherever you go you leave DNA, but you take the soil with you. Clayey soil stuck to a shoe helped solve one of the earliest western soil forensics cases on record. In 1908, the headless corpse of Margarethe Filbert was found near Rockenhausen in Bavaria. Neighbours accused Andreas Schlicher, a suspected poacher. The district attorney, however, felt a need for evidence, and sought the expertise of Georg Popp, a chemist from Frankfurt.

When questioned, Schlicher denied being near the field where Margarethe was murdered. Police found his trousers, rifle and ammunition in a nearby abandoned castle. I can't find any explanation as to why his trousers were there—perhaps the investigators were just as stumped. But by any stretch, misplaced trousers are not enough to convict someone of murder. What got Schlicher into a pickle was his shoes. His wife swore she had, as per her wifely duties, cleaned his dress shoes the night before the murder. These shoes, which should have been clean, had three different layers of soil stuck to the front of the heel. Popp collected soil samples from around

Schlicher's house, the abandoned castle and the murder site, and compared them to soil on the shoes.

On the shoe, the top layer of soil on the heel appeared to be goose droppings and other soil comparable to that around Schlicher's home. The second layer contained fragments of red sandstone like those in the area where the body was found. The bottom layer contained coal dust, cement and other materials like those present where Schlicher's gun and clothes were found. The layers represented a time sequence. Schlicher had been at his house, then the murder site, then the abandoned castle. In his testimony, Schlicher said he had walked only in his fields on the day of the crime—fields that contained quartz, which was not found on his shoes. The soil evidence did not prove he committed the murder. It provided persuasive circumstantial evidence about his movements— and it threw his alibi into question.

What makes soil particularly useful for making or breaking alibis is its diversity.

'There's no such thing as a match,'* Rob says, meaning that no two soil samples are identical, even if they are collected from the same hole. Soil changes both vertically and laterally. There's always a slight variation in a mineral, a bit of a plant root, a worm you interrupted while collecting the sample, and the billions of microorganisms busy keeping the world running. Like us, soil has a fingerprint, but it's a little

---

* In forensic soil science the term 'match' is not permitted to be used.

blurrier. So, no, you will never have two samples with exactly the same properties, but you can be 'highly confident'—as with the soil from Matthew Holding's shovel and the sample from Oakbank quarry—that two soil samples have come from the sample place.

Variation in soil minerals, textures and even colour are not often obvious to the untrained eye. What this means for criminals is that different soil types will counter a false story. Sand from one beach tends to look an awful lot like sand from another beach, even to a trained eye. But under the microscope, differences in grain size and minerals can be crucial—as they were in implicating Carly Ryan's killer.

In 2007, fifteen-year-old Carly Ryan told her mother she was going to stay at a friend's house. The next morning, she was found floating facedown seventy-five kilometres away at Port Elliot in South Australia. Although the main suspect, Garry Newman, was over eight hundred kilometres away from Port Elliot when the police found him, the sand on his shoes compared very strongly with the sand on the beach where Carly was killed. A jury found Newman guilty of murder and sentenced him to life imprisonment with a non-parole term of twenty-nine years.

Scottish forensic soil scientist Professor Lorna Dawson (also adviser to crime writer Ann Cleeves and the very popular crime show *Vera*) helped put away the World's End killer, Angus Sinclair, by closely comparing the soil on one of his victims' feet to the field in which she was buried. In a 2016 interview with Suzanne Allan for BBC Scotland, she said, 'That allowed the prosecution team to show that the alibi that Sinclair came up with in court at the last minute could not possibly have taken place. His version of what had happened could not have happened.'

People lie. Soil doesn't.

Luckily, at least for prosecutors, human interaction with soil makes an already-diverse medium even more complex. Construction, industry, even the common act of driving leave unique chemical and physical markers in the soil. The soil next to a road will have much higher levels of lead, zinc and cadmium than a soil in a forest. Old battlefields tend to have more calcium, phosphorus, iron and lead. Even fungicides used in agriculture can boost soil zinc levels.

Spherical glass particles from a smokestack were strong evidence in convicting Susan Galloway's killer. Susan had already been reported missing when a sharp-eyed hiker spotted her ID beside the Missouri River in Montana. Police soon found her car on the side of a cliff. Susan's body was in the boot, but investigators believed she'd been murdered elsewhere, and police dogs sniffed out a local laneway where someone had tried to dig a hole: Susan's blood was mixed in with the loose soil. The main suspect was Susan's fiancé, Craig Smith. Although Smith had cleaned his sneakers, the forensic geologist managed to collect fifty milligrams of soil (that's 0.05 grams) from the shoes. The soil on Smith's shoes and the soil in the laneway both contained distinctive spherical glass particles pumped into the air over the course of fifty years by a decommissioned copper smelter located nearby. Smith's shoes also had traces of the same soil as the laneway. Combined with the other evidence the police had—changing stories, a mysterious cut and glass-bottle fragments—Smith was convicted and sentenced to one hundred years in prison.

Comparisons

At the heart of forensic soil science is the ability to compare samples. When soil science is involved, investigators are usually trying to find where something or someone was buried, or to track a suspect's movements. The first and most obvious comparison is colour. Colour is a good way to start ruling samples—and suspects—out. Notwithstanding the repeated attempts of men on construction sites to educate me about soil—'it's brown'—there are lots of browns. There are, in fact, nearly seventy types of 'brown' in the Munsell soil colour book, and that's before we get to the reds, yellows, greens and greys. Colour can tell you so much more than 'brown': what minerals are present, for example, or whether the soil has come from a wet or a dry area. Black soil often has high levels of organic matter; red, orange and yellow usually mean different types of iron minerals are present, and white soil may contain lime or be leached (all the colour has washed out into lower layers). Brighter colours mean the soil has been more oxygenated. Dull colours like grey or drab green mean the soil is anaerobic and is low in oxygen, usually in a wet or swampy area. If your shovel has grey clay stuck to it and your suspect has yellow sand on their shoes (and an alibi), start looking at other suspects.

Perhaps the earliest documented case of a forensic comparison of soils was in 1856, in Prussia (modern-day Germany). A train began its journey containing a barrel of silver coins, but when it arrived at the destination the coins were gone and the barrel was full of sand. Professor Christian Ehrenberg, a zoologist and geologist, collected samples from each railway station along the route. Studying the samples under a light microscope, only one station had sand that was the same colour and shape to the sand in the barrel. The police deduced that the only person who could have made the substitution was one of the few employees who worked at that station.

If you're wondering why the thief used local sand, it was probably a matter of convenience. Soil is heavy. Very heavy. One litre of soil weighs about 1.65 kilograms (or one gallon weighs about 13 to 14 pounds). It's hard work to move all that sand about, especially when you're on a tight schedule. Soil has been used in many substitution cases over the years, sending forensic soil scientists and geologists scurrying across the globe to track down missing shipments of cigarettes, gold and weapons. Most of the time the substitution happens close to the soil or sand source.

Forensic geologist Fred Nagle ended up in Brazil when what was supposed to be a two-million-dollar shipment of perfume arrived in Miami as several tonnes of sand. The perfume was supposed to leave Paraguay, travel across Brazil by truck, hop on a ship to Rio and Salvador, then end up in Miami. But somewhere along the way, someone had done the old switcheroo, and no one's insurance company wanted to pay. The sand was fine, brown and rich in quartz. Following the trail and collecting and comparing samples along the way, Nagle traced the shipment's route back to Rio and through Brazil, but with no luck. He couldn't find any sand that matched what was in the shipment. He decided to conclude his travels with a side-trip to see the new hydroelectric dam on the border of Brazil and Paraguay. Noting the brownish sand used in construction, Nagle was told that some of it came from the Paraná River. On analysis, sand from the river was found to be very similar to that in the perfume substitution, revealing that the perfume hadn't made it much past its Paraguayan starting point. No wonder the substitution wasn't picked up along the way—the sand had been there all along.

X-rays, laser beams and $10,000 a day

Different minerals are critical in distinguishing between soil samples. In soil, the minerals are the 'soil' bit that leaves boot prints and makes mud pies. They come from millions of years of the weathering that turns rock into soil. There are over five thousand recognised minerals in different quantities and combinations across the globe. Some minerals, called allotropes, are quite distinct substances made from the same element. To take the best-known example, graphite and diamond are both allotropes of carbon—it's the way those carbon atoms are joined together that makes the difference between a pencil and a potential engagement. In a diamond, each carbon atom is joined to another in four directions, making a strong tetra-hedral shape. In graphite, each carbon atom is only bonded to three other carbon atoms, making a weaker structure.

The X-ray diffraction (XRD) machine is a device that looks at mineral structures to tell us what minerals are in a soil sample. Peter Self, a clay mineralogist who works at CSIRO, is the specialist who does some of the XRD work for the forensic soil lab in Adelaide (Mark Raven does much of the forensic XRD work but was away on the day I visited). Rob Fitzpatrick and I surprised Peter in the lab while he was in the middle of sample preparation, leaning on a bench, mortar and pestle in hand like a modern-day apothecary, grinding a sample down to a fine powder. No matter how advanced you are in your career, every soil scientist gets their hands dirty.

The XRD is a rectangular box with a small arm in the middle that holds the powdered soil sample. The arm rotates very fast while the machine fires X-rays at the sample and measures the intensity of the X-rays scattered from the sample over a range of scattering angles. The scattering process of X-rays from the sample is known as diffraction. When the X-ray beam hits the sample it's diffracted into a specific angular pattern—think of the patterns light makes when it hits a crystal or gemstone in jewellery—and it's like a signa-ture. Because every mineral has a different structure, every mineral diffracts light in a different way.

The XRD measures the different diffraction patterns, does some complicated calculations, then makes a best guess as to what minerals are in the soil sample. The results look a bit like an elevation drawing of a hiking route...if you were going to hike the Trango Towers in Pakistan. Each mineral has unique, characteristic peaks. Peter has been studying minerals for long enough that he can almost read the output the way a musician reads a score.

'But you always check,' he says. 'Trusting the "black box" results in errors. The computer goes a long way to helping by throwing up what it thinks the minerals are, but it's never completely right.' Plus, the more complicated the sample, the harder it is to characterise the different minerals. Most soil samples appear somewhat similar because they usually contain quartz and feldspars, the most common minerals in the Earth's crust.

Although there are databases that contain the different patterns minerals make, the databases aren't complete. And with forensic cases, silly errors can have huge implications, like in the famous Chamberlain case* where the 'foetal haemoglobin' (baby's blood) found in the car turned out to be a sound deadener sprayed on when the car was manufactured.

Peter has a sample running through the XRD when we arrive. As he shows me what the computer has suggested, Rob

---

* In August 1980, two-month-old Azaria Chamberlain was taken by a dingo while her family were on a camping trip near Uluru. With no body and conflicting evidence, her mother Lindy was convicted of Azaria's murder. However, in 1986 during the search for a missing British hiker, searchers uncovered Azaria's matinee jacket. After further investigations, Lindy's conviction was quashed. She had been telling the truth.

paces behind us, making suggestions:

'Kaolinite.'

'Nope,' Peter says, tapping the keyboard.

'Oh, cristobalite!' Rob bounces gently on the balls of his feet. He is not a still man; he hasn't stopped moving since I met him at the bus stop a few hours ago. 'Rutile, anatase— that little peak there—hematite...isn't there smectite in there?'

Peter frowns at the screen.

I ask if they lay bets on this. They don't, but there is a global competition for this sort of work. The Reynolds Cup is named after Bob Reynolds, who pioneered using XRD for soil analysis and who, despite enjoying the challenge of testing his mushroom identification skills by eating the mushrooms, lived to be nearly eighty. The Reynolds Cup is the soil scientists' equivalent of blind wine tasting. Contestants receive a sample of manufactured soil mixed from different clays, and have a few months to identify the type and quantity of minerals in the sample.

'You can do whatever you like to characterise it,' Peter says. 'Grind, kick, smell.' All the contestants use XRD, though. The CSIRO team won the cup in 2010 with a perfect score. Peter shows me a photo of the shiny trophy they got to keep for a year. 'Plus, the winner gets to prepare the samples next time.'

'So it's actually more work,' Rob chimes in with a grin.

Peter doesn't seem to think this is a bad thing.

—

One particularly useful aspect of the XRD is that it can analyse tiny amounts of soil. Soil is like DNA in this respect: you can't see it, the lab doesn't need much, and it can be a criminal's undoing. Craig Smith had cleaned his sneakers—but he couldn't see the fifty milligrams of soil that forensic geologist Jack Wehrenberg used to implicate him in the murder of his fiancée, Susan Galloway. The CAFSS can work with half a milligram. But even if there's less, another piece of equipment can help. The Synchrotron in Melbourne can characterise the minerals in individual grains of clayey soil.

From the outside, the Synchrotron looks like a roofed football stadium, and isn't much smaller. Inside is a network of tunnels and technology that produce beams of light more than one million times brighter than the sun. X-rays and infrared radiation are channelled through long pipelines into different pieces of equipment. Costing ten thousand dollars a day to use, it only gets rolled out in high-profile cases, like the murders of Louise Bell and Corryn Rayney.

Corryn Rayney was a registrar in the West Australian Supreme Court who left her home in 2007 to go to a bootscooting class and never returned. Nine days later her body was found about ten kilometres away in Kings Park. Her husband, a barrister, was an unlikely suspect, but ended up being the main one. Perth is a very sandy place and is one of the few cities in Australia with limited soil mineralogical variation: the soil doesn't change much. I'm not sure Professor Ehrenberg could have located the missing silver coins back in 1856 if they had been pinched from a Perth railway. In this case, the soil itself wasn't much use in trying to track Corryn's movements. The critical evidence was tiny red particles on her body, in her bra and in her car. These particles were not in the soil near her bootscooting class, or in the grave. But there were red bricks on the driveway, footpath and pergola at her house.

These bricks were recycled from housing built before 1940, when lime mortar was used instead of the stronger Portland cement used now. The Western Australia Police Force collected samples from all the different brick pavers in the driveway at her house and gave them to the CAFSS. The CAFSS team spent two years painstakingly linking the tiny brick particles to the old brick pavers in the front yard of her house, even to one specific type of brick. Part of the process required using the Synchrotron, which is far more sensitive than laboratory XRD systems and can identify the minerals if there are only a few sub-millimetre particles. Because the Synchrotron hadn't been used to identify soil minerals before, the defence could easily question the validity of their results. One newspaper called it 'junk science'. Rob now displays this article when he gives a lecture, as two days later his team's work became a major piece of evidence, and their method is now used throughout the world.

In a trial before a judge only, the XRD evidence (along with other forensic evidence) was accepted as proof that Corryn Rayney had made it home from her bootscooting class and was probably attacked in her front yard and not at Kings Park where her body was buried.

Rob was not permitted to provide evidence during the trial as to how the brick particles could have ended up embedded in the fabric of Rayney's bra, because CAFSS had not previously conducted transference experiments. Eighteen hundred bras later and an explanation to CSIRO as to why a man who wouldn't fill an A-cup needed so much lingerie, the CAFSS had conducted enough transference experiments* to publish their method in forensic science journals—which meant Rob could give evidence during the Louise Bell trial.

One summer night in 1983, Louise Bell was abducted from her home, believed to have been enticed outdoors by someone she knew. The critical piece of evidence was her pyjama top, which was mysteriously left on the front lawn of a neighbour's house two months after Louise was abducted. Nearly thirty years later, DNA analysis on the garment revealed only a weak profile—not enough to identify anyone. At the time of Louise's disappearance, soil samples on the pyjama top led the police to an estuary at Noarlunga in South Australia, but they could find no trace of Louise. Although the police collected soil samples during the initial investigation, they had unwittingly collected acid sulfate soils, the one type of soil that will change very quickly if disturbed. Acid sulfate soils are found in low-lying areas along the coast. They sit quite happily under water, but if they are exposed to oxygen—say, by a police officer collecting a sample—they undergo a rapid

---

* PhD student Kathleen Murray conducted a series of transference experiments by repeatedly pushing bras into mud in different ways and dragging them in different directions.

change such that the iron sulfide (pyrite or fool's gold) quickly oxidises to create sulfuric acid and the pH drops down to 2 or less (think battery acid). They also change colour from gluggy green and grey to various shades of brown with yellow and orange mottles.

Over twenty years later the soil samples the police had dutifully collected weren't much use. The pyjama top, however, had stains caused by tiny soil particles embedded between the fibres. There was no way to get the particles out to analyse with the XRD. Instead, the CSIRO team cut little swatches a couple of centimetres square from the soil-stained pyjama top and took them to the Synchrotron, which was able to identify the pyrite and clay minerals while still embedded in the fabric. These minerals compared closely with the soil at the Onkaparinga estuary in Noarlunga, which went some way towards explaining what had happened to Louise, even though she has never been found. Again, the team did transference experiments to work out how small particles of soil became impregnated in gaps between the fibres of Louise's pyjama top. They shook swatches of the pyjama top with various types of soil samples in water, the results suggesting Louise was pushed into the estuary mud.

Dieter Pfenning was charged with Louise's murder in 2013, and found guilty in 2016. The soil forensic evidence supported other circumstantial evidence that ultimately led to the guilty verdict.

Beyond the science

Forensic soil science, collecting samples and doing lab analysis is only part of the job—at least for Rob. Going to court to give evidence can take up half his time, and it's sometimes the most difficult part.

'Is it hard sitting near the accused?' I ask.

For the first time in four hours I have to strain to hear his answer. And he is quite still.

'In Adelaide, I could almost touch him,' Rob says. I don't know which criminal he is referring to, and I don't ask.

Giving evidence means presenting facts, no emotion allowed. This alone is difficult enough, and a day in court is made much worse by the personal and professional attacks from defence barristers looking to discredit him. One day, after a particularly brutal session, Rob decided he'd had enough. He became a scientist because he liked understanding how things worked, not to be attacked in court. No one else in the team wanted to do the courtroom part.

'I even bought him a suit,' Rob says, lamenting his failed attempt to get another team member into the courtroom. But some advice from a forensic scientist colleague in Melbourne has kept Rob in the stand. 'He told me to learn to enjoy the theatre of it. Stick to your evidence and see the rest as a show that the barristers perform, rather than a personal affront.'

I considered training in forensic soil science until I learned how much time you need to spend in court. Having done jury duty at the Supreme Court and seen what the defence barristers do to the witnesses, I decided to keep to the quiet life on the end of a shovel (and pen).

We can, however, thank those defence barristers for refining the Australian Guidelines for Conducting Criminal and Environmental Soil Forensics Investigations into what they are today. Thirty years ago there weren't guidelines on conducting forensic soil investigations. But every procedure needs a method that can be analysed in court, so, as Rob got called in on more and more cases, he started writing it. Every time something in the manual is picked apart in court, Rob updates it based on what the defence barrister has zeroed in on—whether a typo, a misplaced word or an unnecessary sentence. In this way, criminals have inadvertently paid some of the sharpest and most expensive minds in the country to refine the very manual that helps catch criminals to begin with.

'So it's not all bad,' Rob says with a shrug.

Lime burials

Start searching the web for advice on how to dispose of an inconvenient in-law and you'll find an alarming number of threads recommending using lime to dissolve the body. Lime and death are old friends. Lime has been found in graves dating back to the Neolithic period, some twelve thousand years ago. There are no texts from the Neolithic period so we don't know exactly why it was used, but we do know that at some stage lime use was entwined with belief. In the Roman period it's related to Christianity, resurrection, and the colour white, for purity. In the Middle Ages lime was a way to cleanse the soul and body of the miasma, a 'bad air' thought to rise from bodies to cause illness or disease. Liming the walls and surfaces of a house was believed to offer similar protection against miasma.

More recently, lime has been used in mass graves, both legitimate and criminal. It used to be a component of standard natural disaster measures to deal quickly with all the bodies: as a disinfectant, to slow bacterial growth and to reduce odours that would attract scavengers.

You can find lime in fictional murders, in real murders and even in poetry. In 'The Ballad of Reading Gaol' (1897) Oscar Wilde wrote: *And all the while the burning lime / Eats flesh and bone away.*

There's just one problem. It doesn't work.

Eline Schotsmans is a postdoctoral research fellow with the University of Wollongong and University of Bordeaux, and a world expert on lime burials. Like Rob Fitzpatrick, Eline's career took a different path after she received a call from the police.

'We're doing a search. The perpetrator says he's covered the body with lime. Can you tell us what it does? Does it accelerate decay, slow it down or do nothing?'

Eline didn't know. When she started to research it, she found contradictions. Some said lime desiccates the body— dries it out—and I've not thought about desiccated coconut in the same way since I learned this; some said it accelerates decay. In short, despite a widespread popular belief that lime would make a body disappear faster, no one really knew.

So Eline decided to find out. She picked up nine pig carcasses (as analogues for human corpses) and buried them in an acidic sandy-loam soil in Belgium. Three without lime, three with hydrated lime and three with quicklime. Another three carcasses were observed in the laboratory for two and a half months with the same set-up: without lime, with hydrated lime and with quicklime.

In this experimental study Eline found that only quicklime caused an initial acceleration in decomposition because quicklime makes the temperature increase when it is mixed with moisture. But in the long term, both quicklime and hydrated lime hardened to form a barrier between the body and the soil—like a cast—and decomposition slowed. After six months, the limed pigs were decomposing at a much slower rate than the un-limed pigs. By seventeen months the limed pigs were slightly better preserved than the unlimed pigs, and by forty-two months only skeletons remained. What this means is that liming a body won't make it disappear any faster.

And the smell? Lime might initially keep it trapped inside the cast-like barrier that forms. But when there's a crack or a small hole in the cast, the smell that gets out is more intense: it's concentrated and sticks to the lime.

'When I did the pig experiments, the pig without lime was smelly, but after a while it went away. But with the quicklime, once you open the cast, the smell is quite...ugh. It all comes out.' Eline has been speaking with ease about death and decay for the last hour, but at this memory she wrinkles her nose.

Despite what we know now, the idea that lime will deal quickly with an inconvenient corpse is pervasive. There are regular cases. The body of former Serbian president Ivan Stambolić was found in a shallow lime-covered grave. In 2002 a Belgian woman was convicted of murder after she buried her drugged husband (still alive) in the backyard and covered him with lime. In another case the victim was buried facedown in lime, presumably to help destroy their face faster, but by the time they were found the lime had formed a death mask, capturing their facial features. I have two theories as to why the myth survives. First, quicklime is very alkaline: if you put in on your skin it will burn, which might encourage a belief that it works similarly to strong acid, which really can dissolve skin. Indeed, in one case where the Italian mafia executed two victims then completely covered them in lime, when the pathologists got to them forty-eight hours later the victims had third-degree burns on their faces.

Second, if you use lime in acid soil, as all good farmers do, it raises the pH to something more amenable for most crop growth. This makes the soil microbes very happy, and they start breaking down organic matter faster. As people are basically walking, talking organic matter (living people, anyway), there's a kind of logic in there.

Lime won't make a body decompose faster, but it might just improve the surrounding soil. In Eline's experiments, the lime bumped the pH up to from 4 closer to 7, something better for plant growth. It made me wonder…if you put two corpses in the ground, one in an acid soil and one in an alkaline soil, both with lime, then did an aerial survey, could you spot the difference? Would there be greener patches in the acid soil because the lime has made the soil better for plant growth, and the microbes, happy with their feast of a corpse, make lush green grass above? Eline assures me she has lovely pictures of very happy grass growing above the limed pigs.

Because lime is so often used in clandestine graves, I ask Eline if finding white powder in a burial is an indication of foul play.

'It's usually the other way around,' Eline says.

Stumbling across a body is far less common than *Midsomer Murders* would have you believe. What tends to happen is that someone confesses to a murder, then they and the police go looking for the (lime-covered) body. Because most murders are impulsive and the perpetrators aren't in a clear frame of mind when they dispose of the body, many killers struggle to find the graves again. They think they walked two kilometres into the woods, but only walked a hundred metres, or vice versa. Lime is one of the clues you've found the right corpse.

When you have a fascination with the macabre, it's hard not to ask the question most of us have had at some stage. 'If you killed someone...how would you hide the body?'

After years in this kind of work, surely Eline and Rob would know the best method. The answer from both of them?

'Just don't do it. You'll get caught.'

'But if I did commit a murder,' Rob muses, 'I know exactly which defence lawyer to go to, and I know they'll get me off.'

Wine

SOIL AND ROMANCE are not natural bedfellows. Wine is one of the few fields (or vineyards) where they seem to meet. Only among wine people will you find someone who has never been on the useful end of a shovel getting poetic about soil. For several months while researching this chapter, I spent an unhealthy amount of time in liquor shops reading the back-labels. Many described the flavours in the wine—oak, pepper, leather, spice, grapefruit, cashew…others told stories of how the vineyards were founded, and one delightful series from Hesketh Wines told stories about mythical creatures that once roamed the land.

But some labels waxed lyrical (as lyrical as you can be in fifty or so words) about the soils on which the grapes grew. I read about ancient soils, volcanic soils and deep, rich, clayey soils that gave the wine such a spectacular flavour that if I parted with enough money I, too, could taste the volcano. Here my suspicious nature kicked into overdrive. What alerted me to the idea that some labels might be pushing the boundary of hard reality was the recurring mention of 'rich, red volcanic soils' in the Lower Hunter Valley, a few hours north-east of Sydney. I have made multiple trips to the Hunter wine region over the years, including trips to specifically look at soil, and volcanic soil never made it into the discussion. But I am not a geologist or volcanologist, and whenever I tried to learn more about this volcanic soil I was directed to the same person: John Davis.

John is a geologist from the mining industry who knows more than a thing or two about the soil in the Hunter. In the 1980s he literally struck oil and, after a glamorous few years in South America and Asia, did what many professionals with heavy pockets do—he bought a winery. John has had years to get to know the Hunter, so I ask him about rich, red volcanic soils in the region.

'It's a load of bullshit,' he grumbles. John has some of these soils at one of his vineyards: Tallavera Grove, along the edge of Mount View. They are indeed very red. 'Rich' is debatable, but they are definitely not volcanic. The underlying rock, an impressive contrasting white, is limestone, littered with Bryozoan fossils washed in about 280 million years ago. Nearly 300 million years ago, the Hunter Valley was under the sea. For 50 million years, the ocean washed in sediments, while sea creatures, algae and coral used calcium carbonate in the water to make their shells. When creatures died they sank to the sea floor, buried by the sediments that eventually hardened into limestone rock.

You don't need to look that closely to see the fossils and the odd scalloped shell embedded in the rock. Yet the myth that the red soil on top of the rock is volcanic persists. John tells me the story of an (ex-)vineyard manager who raked up all the fossil-filled rocks from a paddock, put them in a neat pile in the corner, then wrote the blurb about the wine being grown on volcanic soil.

'I walked the manager over to the pile and asked, "How did these animals live in a volcano?"' Lava is typically between six and twelve hundred degrees Celsius: not far off cremation temperature, at which all the skin, muscle and fat are vaporised and bones are turned into ash. It is not, in other words, the right medium to make fossils.

'I've even done a demonstration, putting hydrochloric acid on the rock and watching it fizz,' John says with a shake of his head. (Because shells are made of calcium carbonate, they are alkaline. Putting an acid on them makes them fizz—like when you mix baking soda with vinegar.) 'But it doesn't help if someone really wants to believe the rock is volcanic.' It is human nature to believe what you want to believe, regardless of the evidence fizzing in front of your face.

To illustrate his point, John takes me to a geology display at the cellar door—something he put together himself. The display is an assortment of rocks from the property, full of shells and fossils, and large posters explaining the geological history of the winery, the fossils and the soil. He spends twenty minutes happily pointing out different types of fossils and rocks, and explaining geological units and how they formed; there's a poster, too, for when he can't be there to go through it in person. I am not surprised, even though posters like this are usually only made by university students under duress (though as someone who has spent too much time at university, I do appreciate a good poster). You can take the man out of geology...

The persistence of the volcanic soil theme is probably down to a mixture of marketing and geological nuance. Volcanoes are sexy, associated with richness and fertility, whereas fossils conjure images of dusty museums and grumpy geologists reminiscing about the good old days of the Permian. And to say there is nothing volcanic whatsoever in the Hunter Valley is untrue. There was volcanic activity in the wider region during the Carboniferous, about 300 million years ago. The extinct volcano Mount Barrington lies nearly a hundred kilometres to the north, and the Liverpool Range volcanic province is about a hundred and thirty kilometres north-east. But we can be very certain that none of the fossils we are looking at is a previously undiscovered species of volcano-dwelling invertebrate.* That rich, red soil was not ejected from a volcano in a spectacular display of molten rock and pyroclastic ash. It was weathered from limestone.

—

* Although there are 'volcano snails' that live in underwater vents, surviving temperatures up to 400°C.

In fact, limestone-derived soils are some of the best wine-growing soils in the world. *Terra rossa* (loosely translated as 'red earth') soils produce some of the finest wine in France. They're the same soil John has at his vineyard, and are so coveted that on Australia's Limestone Coast they were part of a billion-dollar boundary dispute.

The Coonawarra wine region lies nearly four hundred kilometres south-east of Adelaide. Within the region is a long, thin stretch of *terra rossa* soil known as 'the cigar'. It is a unique selling point that boosts the marketing value of the whole region. Wineries display soil cores on the counter when tasting parties come through. Bright red crumbles over stark white limestone: much more exciting than plain old brown.

Back in the day, vineyards used the price-boosting Coonawarra label widely and joyfully, but then a combination of labelling laws and international treaty obligations changed things: all wine regions in Australia now needed to be clearly delineated and defined. Which meant drawing lines on a map. For a wine to bear the word 'Coonawarra' on its label, it had to be within the newly defined Coonawarra Wine Region—and some businesses that had been producing for years as Coonawarra wineries found themselves outside the boundary. People were unhappy, appeals were made, the courts got involved.

In the red corner, representing those inside the new boundary, was Rob Fitzpatrick, our crime-fighting forensic soil scientist (Crime chapter). In the blue corner was his colleague and friend, the state soil surveyor. Surely two soil experts should be able to settle the case on the facts, quickly and quietly?

Nearly ten years and almost a billion dollars later, the boundary largely remained where it had been set, but not because of any major disagreement about soil types. Part of the case required bringing an excavator in to dig pits (big holes in the ground where soil scientists hang out, peering at the wall of soil in front of them and occasionally making excited notes). Because this was a land dispute there was a field tour for all parties, lawyers and judges included, to also go out and look at the soil. At one point Rob jumped into one of the pits to explain what he could see,* at which point the opposition lawyer threw a minor tantrum and ordered him out—but without the judge knowing. Rob obliged and nothing more was said about the matter, until the very end of the case.

'I was ninety per cent certain the opposition would win,' Rob says. 'Their lawyer was on his game, very convincing.'

A good barrister is often more powerful than evidence but this lawyer, the same one who had ordered Rob out of the soil pit, got overexcited. Instead of going for a humble kill, he wanted to shoot his opponents down in flames. 'And he tried to discredit me,' Rob says with a grin.

How did that go? Not well.

The lawyer claimed Rob had no evidence whatsoever of

---

* If someone is pretending to be a soil scientist, the best way to test them is to show them a soil pit. If they don't climb in, you've got an imposter on your hands.

any special soil properties within the Coonawarra boundary, so Rob started to outline his evidence. He took his time: 'I was very slow, going through the papers, explaining the terms and the soil properties. The opposition's lawyer was getting frustrated. He kept trying to stop me and the judge could see that, so she says, "Doctor Fitzpatrick, why didn't you explain all this while we were doing the tour?"'

Rob answered honestly. 'I wasn't allowed to. I jumped in one of the pits and I was told to get out.'

The judge slammed her gavel, accused the opposition of perverting the court of justice and wasting obscene amounts of money, declared the court closed and stormed out. Her direction to the two barristers was to go away and sort it out.

About one month after the case finished, when Rob and his colleague on the other side were finally allowed to speak after years of forced silence, they went to the pub and hashed it out. They agreed that there were patches outside the boundary that could have been included, and some sections inside that should have been excluded. The boundary was eventually amended, but not as much as some people had hoped—and if the two soil scientists been allowed to speak up-front there might have been a much quicker, cheaper consensus.

Rob still occasionally gets sent bottles of wine and asked to edit the labels. 'I need to have quite a few tastes to be sure of my assessment,' he assures me, and I can't fault his scientific rigour.

The taste of place

We shouldn't need billion-dollar lawsuits to value soil. From an agronomic perspective, the soil in some areas outside the Coonawarra boundary wasn't much 'worse', especially considering that different grape varieties prefer different soil types. For example, semillon does well on sandy soil, and some reds do better on loamier soils. And being on the 'right' side of a line on a map doesn't guarantee a superior wine.

What the Coonawarra boundary dispute did show was how valuable soil is to marketing. In an increasingly competitive market, products need a 'point of difference', like attractive *terra rossa* soil. The point of difference many vineyards are using now is *terroir*, a French word that means roughly the 'taste of place'. Terroir encompasses the environmental conditions that make a place what it is—aspect, slope, light, climate, biology and soil. These elements work together to create specific flavours that are characteristic to a region. Terroir is why a cabernet sauvignon from the Médoc in Bordeaux tastes different from a cabernet sauvignon from the Napa Valley in California, the cooler French climate being just one piece of the puzzle. Or why a pinot noir from one side of a hill in Otago, New Zealand, can taste different from a pinot noir on the other.

Terroir is a centuries-old concept, now refined and tightly controlled in France—though even they still move their borders. In 2017, the Champagne region was expanded for the first time in ninety years. Far more than just lines on a map, there are more than three hundred regions, *appellations d'origine contrôlées,* that have specific rules controlling every part of winemaking from pruning to pressing. This is why bubbly produced outside the Champagne region, or bubbly that doesn't follow every rule for Champagne production (such as secondary fermentation in the bottle) must, by law, be called something else, like sparkling.

Soil is only one part of terroir, yet it has become the poster child for the taste of place. As we've seen, you can find soil cores at cellar doors; you can also go to 'soil tastings' where you literally smell and taste wineglasses full of soil, and enthusiastic wine journalists say things like 'soils trump climate' when it comes to influencing the taste of wine.

Is there any truth to it? Paddocks are complex places, and I've dug more than enough holes to know that soil can vary from metre to metre. Does soil really impart a distinct flavour to wine?

'Absolutely,' says Gwyn Olsen, winemaker and assessor. 'You can definitely taste the difference between a shiraz grown on a sandy loam, or clay, or rock.'

It's easy to be sceptical about wine 'experts', especially when you read about experiments in which master tasters are served the same wine three times in a row without realising. Or the time when PhD student Frédéric Brochet fooled multiple oenology students by dyeing a white wine red.

But I feel compelled to believe Gwyn. She certainly has the credentials. Dux of the Australian Wine Research Institute (AWRI) tasting exam, subjecting her palate to about five hundred wines over five days, she was able to pick out the repeating wines the examiners had snuck in. She's a Len Evans Scholar (one of twelve people selected each year to be trained as a show judge), and her palate has been statistically verified. In early 2020, when it seemed that most of Australia was on fire and was tainting our wines with 'ashy' or 'burnt' flavours, Gwyn took part in a double-blind smoke assessment taste test. She was the only assessor to get the negative references correct, meaning she could taste which wines didn't have smoke taint. So if she says she can tell what broad soil type a grape was grown on, she probably can. Red dye in a glass probably wouldn't fool her either.

'My husband, a wine buyer, likes to give me glasses of wine and get me to guess where they come from.' Her tone suggests she's quite accurate. 'The Western Australian chardonnays are easy—they use the Gingin chardonnay clone* which I can pick just by smell.'

To taste and smell these differences in wine takes many years of practice, developing a memory bank of all the flavour profiles, and learning the vocabulary to communicate them.

'It sounds fun but it can be an exhausting process,' says Gwyn, 'walking into class at 8am on a Wednesday morning and seeing fifty cabernets lined up, after you spent all Tuesday tasting wine.' And it destroys your teeth. The pH of wine varies from 2.9 to 3.8, which is acidic enough to dissolve tooth

---

* This is a type of vine. Certain vines are selected or bred for certain qualities, such as pest or disease resistance, or flavour. When a vine works well, cuttings are made from the 'mother vine' and grafted onto rootstock to preserve these qualities. The Gingin clone gives Margaret River chardonnays lower yields but unique flavours.

enamel. 'The worst thing I can do, after a day of judging, tasting a hundred and twenty wines, is brush my teeth. I'll just brush the enamel right off.'

While there are flavour differences in wine attributable to different soils, you can't actually taste the soil. For a few years, though, some wine critics believed that you could. About thirty years ago 'minerality' was coming up a lot in tasting notes. 'Mineral-rich', 'mineral notes', 'crushed minerals' or even 'crushed stones' to describe flavour or aroma, or both.

In the literal sense, minerality suggests that the different geological minerals in the soil have worked their way up through the roots and into the grapes, infusing the wine with flavours like granite, chalk or flint. It is a picturesque notion that, as Welsh geologist Alex Maltman so elegantly puts it, harks back to a time before photosynthesis was discovered.* Plants take up nutrients dissolved in water through their roots, but they do not take up pieces of soil or rock. Plants do take up mineral nutrients (or elements) like nitrogen, phosphorus and calcium when they're dissolved in water, but all plants take up the same thirteen or so elements from the soil (and sometimes their leaves: foliar feeding). 'Mineral-rich' soils imparting greater minerality to wine is a dubious concept because all rocks are made of minerals, and nutrient-rich soils

---

* In the 1600s, the prevailing scientific theory was that plants were made from and ate soil. With only four elements to choose from (water, air, earth, fire) it was logical that plants were made from either soil or water. It wasn't until the very late 1700s, after the untimely demise of many mice sealed in jars with and without plants, that photosynthesis was discovered.

tend to make showy vines and crappy grapes.

The clincher is that even if vines could take up geological minerals, over ninety-nine per cent of the known minerals are odourless, and therefore largely flavourless. For something to have flavour* it must have a smell, and to have a smell it needs to contain some sort of volatile chemical that activates sensory cells in the nose. Rocks don't contain volatiles. You can test the interaction out by eating a jellybean with your eyes shut and your nose blocked and see if you can tell what flavour it is. Most people can only taste 'sweet'. If you pick up a rock and lick it, what you are tasting is whatever residue is on the rock. It might be petrichor, the compound that settles on the landscape and is responsible for the 'smell of rain'. Scrub the rock clean and usually it will have no taste, only texture.† Also note that in this respect rocks are like toads: some are poisonous, so it's not a good idea to go around licking them.

One relevant exception, says Robert White, soil-wine specialist and author of numerous books on managing soil in vineyards, is halite. Too much halite (sodium chloride, aka common salt) in a wine gives it a salty taste. For obvious reasons, vineyards want to avoid having too much salt in their wines; from a viticultural perspective, too much salt in the soil kills the plant.

---

* Beyond the basics of sweet, sour, bitter, salt and the more recently accepted umami, which are sensed by the tongue.
† A few minerals do have a taste. Borax has a sweet alkaline taste. Ulexite is alkaline. Sylvite is bitter. Glauberite is described as being bitter and salty. Melanterite has a sweet, astringent, metallic taste. Chalcanthite has a sweet metallic taste and is slightly poisonous. Epsomite is bitter.

Minerality, therefore, is likely another combination of marketing and nuance. Some wines do have earthy flavours; they just don't come directly from the soil. What happens is that flavour or aroma compounds found in the soil are also found in some wines. When Swiss researchers were investigating why toilets smell so bad, they put hydrogen sulfide (H_2S) and dinitrogen (N_2) into an olfactometer, a device that measures the concentration and intensity of different odours. By accident, they found that when H_2S was the only compound in there, after a while it no longer smelled like rotten eggs, but took on a 'flinty' odour akin to the 'cold smell of fireworks'. They then tested two Chasselas wines (a dry Swiss white) that had been identified in blind tasting has having a flinty mineral flavour. The wines with the flinty flavour had more hydrogen disulfide (H_2S_2). Something is happening with sulfur compounds that creates the flinty odour. Flavour science is incredibly complex and there is no clear answer yet, but we can conclude that the flinty compounds haven't been taken up directly from the soil and then survived fermentation. Sulfides can occur naturally during fermentation—perhaps this plays a role in the flinty notes?

Geosmin is another compound that can give wines a mineral flavour. It smells like damp soil, is a strong contributor to the earthy taste of beetroot, and contributes to the earthy smell of petrichor—the smell of rain. In the soil, geosmin is largely produced by *Streptomyces* bacteria. But, again, it doesn't get into wine directly from the soil. It can come from contaminated water and when using grapes that have grey mould. *Botrytis cinerea* (the mould that makes dessert wine) and *Penicillium expansum* (cause of the devastating blue mould in apples) interact in a way that gives higher concentrations of geosmin. Even some *Penicillium* moulds on corks can synthesise geosmin.

So where does that leave us with minerality?

'I think of minerality as...when you drink running water out of a stream and you can almost taste the wet stone it flows over. I attribute minerality as a tangible perception of rock,' says Gwyn. 'Does that sound wanky?'

I assure her it does not, and we agree that minerality is a communication tool, like all the other wine flavour descriptions. You don't expect there to be pieces of grass, cigar or old boot in a wine.

'Mousiness' is a good example,' says Gwyn. Mousiness is caused by various lactic acid bacteria and *Brettanomyces*, and is particularly sneaky because you can't smell it first. 'Of course there are no actual mice in the wine, but when a wine has "mousiness" the perception of the flavour is like licking the bottom of the mouse cage, even though you've (hopefully) never done it.'

Nature or nurture

Winemakers are, first and foremost, chemists. By balancing and managing the hundreds of different compounds and characteristics of grape juice they work to elicit certain flavours, aromas and mouth-feels. Winemakers have an increasingly sophisticated arsenal—yeasts, acids, an assortment of clarifying substances like bentonite clay, colour, tannins, oak flavour, enzymes that remove certain proteins, and so on—to help create the perfect ferment. Which might make you wonder: how much influence does a winemaker have? Can they obliterate, with a dash of this and a splash of that, any flavour that came from the paddock?

'We have a saying in the wine world,' says Gwyn. 'You can't make a silk purse out of a sow's ear.' Or, less delicately, you can't polish a turd.

No winemaker, no matter how talented, can completely wipe out the underlying taste of the grapes. Take, for example, the grapes tainted by smoke from the summer 2019–20 fires. Vines close to a fire take up volatile phenols produced when wood is burnt and store them in the grape sugars. It is not possible to remove those compounds, Gwyn says. 'I can mask them, I can complement them, but I can't remove them.'

Taints aside, how much a winemaker needs to modulate a wine largely depends on the price point. Significantly manipulating flavour is part and parcel of making commercial wines that consumers expect to taste the same every time. Mass-produced brands blend grapes from multiple vineyards, regions and seasons, each batch contributing different flavour that needs to be tempered.

For single-vineyard wines most of the flavour work is done by the vines. 'Single-vineyard wine grapes should be of such high quality that you don't have to do much to it in the winery,' says Gwyn. A good winemaker knows when *not* to touch a wine.

The goal, then, for a single-vineyard wine, is to coax as much flavour as possible into the grapes while you are growing them. And that means starting with the soil.

The acidity grapes develop while growing is a critical contributor to wine quality and flavour. The normal pH range for wine is about 3.4 to 3.8. Any more acidic and it can turn sour like vinegar; too little acid makes a dreary, flat-tasting wine that is prone to spoilage. The impact of soil pH on grape acidity is interesting. Research in Oregon found that as soil pH increased, grape acidity decreased and vice versa. The final products, pinot noirs, were 'brisk' and 'less complex' on high-pH soil, while the acidic soils made rounded, more complex flavours. Potassium can also influence grape acidity: too much potassium in the soil can lower acidity and lead to colour instability in red wines. And, while winemakers can add acids to correct pH, it's better to start with grape juice close to the ideal pH. Manipulating acidity isn't a panacea, and it can exacerbate other problems.

Controlling water and nutrient supply to the vines is also critical for flavour. In this aspect, a little tough love can go a long way. Too much water and food at once, particularly nitrogen, and the vines get lazy, producing big green leaves instead of focusing their attention on growing grapes. Big leaves don't just benefit at the expense of grapes—they can often shade the grapes, which can be problematic for red varieties as it slows colour and flavour development during ripening.

Too little water, on the other hand, and the vine gets overly stressed. The goal is mild panic: convincing the vine it's going to die of hunger and/or thirst without actually harming it, so it prepares for its demise by directing its efforts to producing tasty and colourful fruit that will attract dispersers. Many plants tend to attempt reproduction when stressed—this is why lettuce 'bolts' or makes flowers early when it's too hot or dry.

The soil texture and geology of the vineyard play a huge role in water control. Sandy soils drain fast; dense clays tend to waterlog. Vines need a soil with enough drainage to ensure the soil doesn't get too gluggy (vines don't like wet feet) but not so porous that the vines get parched after only a few days without rain. Irrigation can help, but it gets very expensive very fast and for vines, it's not just the topsoil—about the top fifteen centimetres—that matters. Vine roots can grow many metres into the soil if the conditions are right, and what's happening much deeper is more important.

Take, for example, the gravelly soils of the Médoc in Bordeaux, which look somewhat like moon soil and struggle even to grow grass. But deeper within the soil are shale layers that the vines seek out, growing up to twenty metres deep in the search for food and water. The water drains readily from the sand, keeping the roots well aerated, while the shale layers hold the goodies for the vine to absorb when it needs to.

Timing when the vines get stressed is important too, though it depends on the grape variety. Too little water when the grapes are forming just means fewer grapes. Minor water stress when ripening reds increases colour, aroma and flavour. Chardonnay doesn't benefit from water stress, doing better on deep soil with good water availability throughout the growing season. Sauvignon blanc prefers a constant, though regulated, supply of nitrogen and water. Some research from Oregon suggests that stressing pinot noir after the onset of ripening makes tastier grapes. (By the sounds of it, though, pinot noir is just a pain all round. Having thin skin, it is susceptible to temperature fluctuations, sunburn and disease. It likes sun but not heat, hates wet feet and is low yielding.)

That's not all, of course. Winemaking is a very complex process, a blend of science and art. There are multiple books written solely on vineyard soil management; viticulture and winemaking is a three-year degree, with subjects dedicated just to wine chemistry. The above factors are just some of the many, many aspects of viticulture that the winemaker and vineyard manager juggle—but they're the ones where soil plays a prominent role.

The black box

I began researching this chapter a sceptic, but have now accepted that a) soil can influence grape flavour, b) flavour can make it through to the finished wine, and c) people like Gwyn Olsen, although they're rare, can taste those differences.

The remarkable thing is that neither Gwyn nor the wine scientists can tell you exactly what she's tasting. There's a gaping great hole of scientific knowledge between the vineyard and the bottle. A pinot noir can contain over eight hundred organic compounds that make colour, flavour and aroma—and for many, we don't know which ones were made in the paddock, or why they were made.

Markus Herderich and team from the Australian Wine Research Institute (AWRI) are trying to unravel the mystery. Markus is a food chemist who hails from Germany, and is everything I expect from the stereotype. Tall, efficient, thorough and beyond punctual (he apologised for being only ten minutes early).

'We only know of about four flavour compounds in a finished wine that are made in the field and not created or changed during winemaking,' Markus says. 'There's a lot of work to do.'

Their most famous success so far, even though Markus hints there are other breakthroughs on the way, is discovering what gives some Australian shiraz grapes that distinct 'peppery' flavour and how the landscape affects it. Rotundone, a compound found in peppercorns, is also present in grape skins. Concentrations of rotundone are highest in areas of low light and cooler temperature, so it makes quite a difference whereabouts on a hill the vines are planted. Grapes lower on a slope get less light and have more rotundone; concentrations decrease as you go up the hill towards the light. Concentrations can also vary within a single bunch of grapes. When scientists tested grapes from different vineyards with various levels of sun exposure, the grapes at the top back of a bunch—those that were shaded the most—had the highest concentration of rotundone.

The AWRI team have also worked out why some grapes

develop eucalyptus flavour, which is great in cough lollies but not appreciated by all winemakers. Crush a eucalyptus leaf and you will smell 1,8-cineole, a compound that is also found in mint. Like rotundone, it is affected by landscape, but the relationship is even more obvious. Vineyards surrounded by eucalyptus trees, often used as windbreaks, have more eucalyptus character because tiny fragments of leaves and twigs end up on the grapes and in the ferment. Hand harvesting fruit, a gentler and more precise process, leads to less 1,8-cineole than you get with machine-harvested grapes.

'We need to ask the grapevine,' Markus says. 'No one really understands how or why a grapevine decides to make certain flavour compounds.'

It's a fascinating point that also suggests untangling terroir is not a simple matching game. Identifying which compounds are responsible for which flavours is just the first step. Then we need to understand why. What encourages chardonnay to make thiols that generate a flinty aroma that's sometimes associated with minerality? Is it some soil or landscape factor that encourages methoxypyrazines, the grassy note in some sauvignon blancs?

That vineyards are incredibly complicated environments only multiplies the challenge. Light, heat, temperature and all the microclimates change every hour of every day of every season. One unexpectedly hot day might significantly alter flavour: when a grape shrivels, not only are some flavour compounds concentrated, new ones are made that aren't present if the grape doesn't shrivel. Could we 'undo' this by adding more water, or are those flavour compounds there for good?

As Markus said, there's a lot of work to do.

Piecing it all together will take decades and many millions of dollars. As science advances, new discoveries enter the mix, adding further layers of complexity. At the time of writing this chapter, microbial terroir was very popular. Vineyards are not sterile production systems. Could the billions of microbes shuffling about on the grapes and in the soil cause a plant to make certain compounds, or switch on defence mecha-nisms—in the vine itself or in other microbes—that have an effect on the vine?

Pathogens usually get the limelight, but even they some-times have a happy side effect. Take *Botrytis cinerea*, the rotting fungus responsible for dessert wine. There are a few origin stories from various nations claiming to have made it first, but the consistent narrative is that something happened to delay harvest (bandits, a late messenger, poor time-management skills...) and the grapes started to rot. Waste not want not; the spoiled grapes were fermented anyway and dessert wine was born.

Some vineyards are embracing microbial diversity, using spontaneous or indigenous ferments to elicit the taste of place. Instead of using purchased yeasts with known flavours and characteristics, the winemaker leaves fermentation decisions up to whatever organisms happen to be on the fruit at the time. It makes more interesting and varied flavours, and is something else the AWRI team is investigating.

'You can taste and smell the difference between the microbes from different vineyards,' says Markus.

US researchers have also found that the four major wine-growing regions in California all have different microbiomes. But once again, we mostly don't know which microbes are making what flavours.

I'm certain that one day soon, indigenous ferments will move from novel to normal, but they're likely to stay at the expensive end of the shop. Grape quality is exceedingly important—and the winemaker cedes even more control and takes on much greater risk. Perhaps by the time the second edition of this book is published, I'll have whole sections on soil microbes. But personally I'm in two minds about unravelling it all. The scientist in me would love to identify every flavour compound and know what conditions encouraged them, then play around trying to manipulate flavours; the writer in me enjoys the romance of the unknown.

As for the wine labels, I've largely stopped reading them. Unless any vineyards want a soil scientist to review them, in which case my glass and pen are ready.

Health

ONCE UPON A time, at least a few hundred thousand years ago, some humans started using soil for its health and healing properties. Why we thought to plaster mud on ourselves, let alone eat it, we will never know. Maybe a tribe of Neanderthals saw an animal smearing mud across a wound or a mammoth bathing in mud to cool down; or they just watched their kids playing. Random experimentation presumably spurred most human progress in the early millennia.

Some of it must have worked, at least in part, because soil-based remedies persisted through Ancient Greece, Ancient Egypt and Mesopotamia. Cleopatra used mud masks from the Dead Sea to main maintain her infamous beauty. Peter the Great sent for medicinal water and mud from the Tambukan Lake, and composer Tchaikovsky worked on his opera

Voyevoda while enjoying the restorative mud spa in Haapsalu, Estonia.

When something persists for thousands of years there is usually a good reason. The Ancient Greeks were onto aspirin (isolated from willow bark) 3500 years ago. Morphine is still harvested from poppies, for the simple reason that despite all our advances in technology, poppies make morphine cheaper than people can.

Thus: clay. For some people, historically and today, clay is their coconut oil, their apple cider vinegar, their Windex, their gaffer tape—able to fix everything from toothache to terrorism (though a Polish friend tells me that vodka does all the same things). Bitten by a snake? Get the clay. Skin problem? Clay. Poisoned by a disgruntled servant? You guessed it: clay.

Eating clay for medicinal purposes would make a book on its own, though I touch on it briefly in the antibacterial clay section. The chapter on geophagy covers eating clay for the joy of it. But in the health department clay is more commonly and effectively used for skin issues. Whether mopping up nasty toxins or helping stop a life-threatening haemorrhage, some clays have literally saved lives.

When clay saves the day

'There is some very graphic stuff in here,' says Matt Pepper. 'Is that cool?' It's an afterthought: the first gory picture is already on display.

Fair enough—Matt usually gives presentations like this to

police officers, military medics and paramedics—people who have actively signed up to deal with the graphic stuff. After seven years in the army and fifteen years as a special operations paramedic, he now runs the training department for an emergency and medical supply store.

Bike accidents, gunshot wounds and a whole lot of blood: for the next thirty minutes Matt takes me through photos and videos of various life-threatening injuries. In each example he explains how quickly the person would have died if they had not received quick medical attention. In all his examples, wound packing was the main reason they did not die. Medics pack certain wounds when they can't get a tourniquet on: either it's too far up the limb or, more often, it's in a junction such as the groin or armpit.

'The first thing you need to do is get your fingers in there to stop the bleeding,' Matt says. He demonstrates this later in the carpark, with some fake blood that I pump through a rubber severed thigh, red watery liquid squirting out onto the pavement until Matt's strong hands stem the flow and pack the wound with bandages. Enjoying the challenge, I squeeze the bottle as hard as I can, but not even one drop of fake blood oozes from the deftly packed wound.

The bandages used to pack the wounds are impregnated with kaolin clay because kaolin (among other uses, as we'll see later) promotes blood clotting. While the medic applies pressure to stop the patient from bleeding out, the kaolin initiates clotting by triggering a protein called Factor XII. Triggering the clotting process means clots form faster—and even if it's only a few minutes, that can be the difference between life and death.

Kaolin-impregnated bandages and gauze have been used since about 2008. Earlier versions of products to stop bleeding used granules mixed with zeolite that were poured into open wounds. This did promote clotting, but also triggered an exothermic reaction strong enough to cause second-degree burns. Kaolin works about as well as zeolite but without the burning.

The downside of kaolin is it relies on your intrinsic clotting cascade. 'So if you're hypothermic and are bleeding out,' says Matt, 'your clotting factors are less effective.' Basically your body doesn't have the capacity to clot, so there is not much value in trying to trigger the process. In that case, Matt prefers a different procoagulant extracted from shrimp shells. When this product comes into contact with blood it swells and sticks together, making a 'snotty plug', as Matt puts it, more technically known as a mucoadhesive plug. Because this system works independently of the body's own clotting mechanism it does not matter if the patient is low on clotting factors.

Fuller's earth is another multi-use clay mineral material. Once upon a time it was used in laundries as a way to remove oil and grease from very grubby clothes. Fuller's earth contains mostly bentonite and has a high CEC ('cation exchange capacity'—we'll look at that in greater detail later) which makes it extremely absorbent. It basically means that the same stuff used in kitty litter can also work wonders for decontaminating chemical agents such as nerve agents, blister agents and mustard gas—weapons of chemical warfare that wreak appalling physical damage. Fuller's earth won't remove it from the atmosphere, but if some lands on your skin a quick application of the stuff can trap the agent's particles and prevent them from permeating the skin. The clay is then brushed off—carefully, so it doesn't suspend in the air where it can get into your eyes or mouth and hurt you anyway. Once it has removed the agent, the Fuller's earth needs to be contained and decontaminated.

Antibacterial clay

It was an unanswered email to the Clay Minerals Society (supported by a rousing guilt-trip) that led Lynda Williams, a geochemist specialising in clay minerals, into researching antibacterial clays. The email was from a man requesting information on behalf of his mother, Line Brunet de Courssou, who did not speak English. She had travelled to the African republic of Côte d'Ivoire with her husband, a French diplomat, and had been treating a nasty flesh-eating bacteria called Buruli ulcer using two different types of French green clay.

Buruli ulcer, related to leprosy, is caused by *Mycobacterium ulcerans,* which is thought to be transmitted by mosquitoes. It starts with a sore that grows and spreads, destroying the skin and fat, leaving painful open wounds that do not heal. There were no antibiotics available at the time, and Line, who had grown up in the Loire Valley, had used French green clays on her childhood cuts and bites and found it made them heal more quickly. She tried the local Côte d'Ivoire clays first. They did not help, so she imported some green clay from France at her own expense, and that was working: healing the ulcers in a few months. Line had now gone to the World Health Organization in Geneva to ask for funds to continue this treatment, but they would not support her unless she found scientists who could explain how the clay was healing this infection. Hence the email to the Clay Minerals Society.

None of the clay mineralogists knew, so no one responded. But after a second somewhat accusatory email complaining of American scientists' disinclination to help the poor in Africa—and being a 'bleeding-heart liberal', as Lynda describes herself—she sent a reply. She didn't know anything about the medical applications of clays, she said, but she did have access to electron microscopes. She could take a look at the clays and see if there was anything unusual about them.

'He thought that they were fibrous clays, he called them "nano-scalpels" and he thought that the nano-scalpels were spearing the bacteria and killing them that way. But under the microscope they were near-perfect little hexagonal plates of illite-smectite.' Lynda smiles as she describes the clay, talking about the particles the way someone else might talk about a pet guinea pig.

However the clay was working, it wasn't through physical destruction.

Line had used two different types of green clay on her patients—Argicur and Argiletz. The Argicur was very painful when it went on the wounds. After a few days the wounds turned purple, and Line switched to the Argiletz—the 'milder' clay, as she described it. She changed the poultices every day and in time the skin would granulate and the wounds healed completely. There was some scarring, but not too much.

Intrigued, Lynda began doing some tests on the clays. The first thing she found was that the painful clay, the Argicur, had a pH of 10. This is very alkaline for soil, about the same pH as an antacid tablet. The second clay had a pH of 8 to 9: less alkaline, and 'milder' because it didn't burn so much going on an open wound.

Next, she discovered that only the highly alkaline Argicur was antibacterial, killing various strains of *E. coli*, *Salmonella*, *Staphylococcus* and *Mycobacterium*, while the Argiletz was not. Maybe the Argiletz simply protected the skin and helped it heal, maybe it promoted skin granulation—they didn't know.

Lynda tried to find where in France the clays came from but the supplier has refused to say. However, the benefit of being a soil scientist is that you can make educated guesses. Lynda and some geology mates think the source is volcanic, and from somewhere in the middle of France.

Lynda looked around for other antibacterial clays, and noticed patterns in her findings. Antibacterial clays were blue, green or greyish in colour and had either a low or high pH, below 5 or above about 9. Clays with a more neutral pH were not antibacterial. The colour was a clue that reduced iron—$Fe(II)$, which has formed under low-oxygen conditions—was an important part of the equation. Not the amount—the

French clay only had about four to six per cent iron—but its oxidation state. Its pH was important too, because pH affects the availability of various elements: in a soil with a more neutral pH, around 6 to 7, elements like iron and aluminium are more tightly bound in the soil. When pH drops very low or rises, they can become soluble (which is why aluminium toxicity is a problem for some plants in very acidic soil).

'It turns out that when you add natural water, spring water or tap water to a small amount of the clays, they dissolve. Not completely dissolved like a sugar or a salt, but part of the silicate structure dissolves. If I shake a clay for twenty-four hours with water and I analyse the water chemistry, all of them will have high aluminium, high iron and maybe some calcium or other elements that were in that mineral,' Lynda says.

This led to the idea that the metals—aluminium and reduced iron—were part of the antibacterial process. The high pH meant that these metals were soluble and ready to go into action against the bacteria: the aluminium damages the bacteria's cell walls, then iron can get into the cells, where an oxidation reaction occurs that stresses the proteins and DNA.

'We have done tests to show that there is damage to internal proteins and that there is damage to linear and circular DNA inside the cell. We found one [clay] deposit that killed everything we tested it on, including methicillin-resistant *Staphylococcus aureus* (MRSA).'

This is big. MRSA is a huge is problem in hospitals now because it is an antibiotic-resistant infection.

Lynda took her ten years of research to the Mayo Clinic in Rochester, Minnesota. They approved a small trial on mice— necessary before there can be any human research (unless you're in Côte d'Ivoire). In the trial the backs of the mice are infected with *Staphylococcus aureus* and treated with the clay. The initial results have been, surprisingly, disappointing.

'It hasn't worked so far, so we're trying to figure out why. We have seen the clay work on humans; people (my students) have used this blue clay on their own wounds, which healed quickly.'

Lynda's theory is the moisture and oxygen need to be balanced. The clay poultice needs to stay moist so the metals can be released and work their magic on the bacteria. Oxygen is also needed for the oxidation reaction. Once the iron is within the bacteria there is a series of reactions (called Fenton reactions) between the reduced iron, oxygen and water: these create hydroxyl radicals that damage the bacteria.

The problem with the mice is they won't follow stern instructions to not scratch the clay off their backs, so the researchers have covered the clay, and now believe that the substance they used may not be porous enough. Soon after our chat, Lynda was going to try gauze, a more porous medium— the same stuff that Line used in Côte d'Ivoire.

If this trial works, Lynda's team will aim for a much bigger research program.

'Program directors are excited about this research, but you have to have pilot data to show that it is worth them spending a couple of million dollars on more research.'

Fair enough.

Despite the issues with the mice, the clays remain very promising. Further trials with the Mayo Clinic have found that some clays, including the Oregon blue clay Lynda's team have been working on, have killed every biofilm they had in their infectious-disease research library. (A biofilm is a group of microorganisms that have clumped together to form a 'film' on a surface; a common example is plaque on your teeth.) Biofilms are more difficult to remove and treat as the biofilm sets up its own defences and can block antibiotics. Most recurrent bacterial infections are from biofilms.

There are also many unanswered questions. Understanding the mechanisms—exactly how the clay works against bacteria—is critical. We know that it damages cell walls, we know that it damages proteins, but we don't know why the bacteria let this happen. Bacteria use iron to respire and can defend themselves against Fenton reactions—so why aren't they?

'It has to be something that is common to all these different antibiotic-resistant bacteria because they all die when we put the clay on them,' says Lynda.

And it does seem that the clay, not just the metal content, is crucial. Some researchers tried just putting reduced iron in solution, and that worked for about half an hour. It is too short a time to be effective, whereas when using the clay the process works for as long as a week because the clay operates as a reservoir of reduced iron: it's a time-release capsule.

The goal for Lynda is to understand the mechanisms so that, in time, someone will invent a synthetic clay that can do all the same things, but without having to use mined clay.

'It is highly unlikely that regulatory agencies would approve a natural clay because they are different in every shovelful.'

At the scale these products would be needed, the consistency of synthesised materials would be important. You might find a natural deposit with the right composition and chemistry, but dig for a few weeks and the composition could change, making it less useful or perhaps even harmful. Some mining companies have encountered such an issue when their deposits start to contain lead and arsenic and mining has to cease.

Manufactured clay would also get around the particle-size issue. Nanoclay particles, which occur naturally, are smaller than fifty nanometres and are dangerous on open wounds because they can get into the bloodstream and cause thrombosis of the lungs and brain.

Like Lynda, I hope to see manufactured versions of this potentially wonderful technology on the shelves. It will be safer and more reliable than mined clay, and should not require clay mining to manufacture. And given the sadly pervasive reality that not everyone in the world will have access to any products developed out of this research, Lynda hopes that knowing what properties to look for when seeking an antibacterial clay will allow people anywhere to find natural clays to use in an emergency.

—

Microbes

It's a tough life being a bug in the mud.
—Rob Capon

'My kids—and my grandkids if my kids ever get off their butts and do something about it—will have a lower level of protection against pathogens than I will...and that's not right. Things are supposed to get better with every generation,' says Rob Capon, impatient grandfather-in-waiting and professorial research fellow at the Institute for Molecular Bioscience at the University of Queensland.

Rob has spent his career hunting through the natural world for molecules that could be repurposed for the greater good, whether agricultural benefit, animal health or human health.

Antibiotic resistance is one of the most pressing health issues of our time. In the US, nearly three million people contract an antibiotic-resistant infection each year, resulting in more than thirty-five thousand deaths. Globally, deaths are estimated at more than a million per year and, in Australia, deaths due to antibiotic-resistant bacteria are higher today than a decade ago. Because microbes are crafty creatures that adapt to their environment, antibiotic resistance occurs naturally; however, misuse and overuse of antibiotics—in both people and animals—is making the problem worse. Medicine development is not keeping up with rates of antibiotic resistance, so scientists like Rob Capon are turning to nature and digging holes across the globe in search of the next wonder drug.

Scratching about in the soil in the pursuit of useful

compounds is not a new idea. Scientists have been busily smearing agar-filled Petri dishes with soil for over a hundred years, trying to ascertain what microbes are hanging out in there and what their purpose, and uses, might be. The research has been rather successful. About forty per cent of prescription drugs have their origin in soil, and in the decade from the mid-1980s to the mid-'90s, sixty per cent of newly approved drugs came from soil. If you have ever taken streptomycin,* an antibiotic used to treat skin, eye and ear infections, and the first effective drug against tuberculosis, you can thank actinomycetes. This group of tiny soil-dwelling bacteria act a bit like fungi by forming threads, and are responsible for over half of human antibiotics, as well as giving soil its 'earthy' smell by making geosmin, the compound responsible for the scent. Tetracyclines, used to treat acne, malaria and some tropical infections, are also derived from actinomycetes.

The biggest-selling drugs in history, statins (cholesterol-lowering drugs), are derived from soil microbes: *Aspergillus terreus* is a common soil-dwelling fungus that produces a

---

* Streptomycin's discovery is shrouded in controversy. In one version, in 1943 Albert Schatz, an 'intense, skinny postgraduate' at Rutgers University, spent months going through soil samples looking for something to work against gram-negative bacteria including the microbe responsible for tuberculosis. He found one but Schatz's supervisor, Selman Waksman, took the credit, got the Nobel Prize and made a significant amount of money— millions of dollars today—from a patent. Waksman convinced Schatz to sign over his patent to the university, but kept his own. Schatz eventually sued and won, but with serious damage to his career. In another spin on the tale, Schatz only made a minor contribution by doing basic laboratory work. In both stories, other laboratory technicians who worked on streptomycin, including Elizabeth Bugie and H. Christine Reilly, are often left out altogether.

compound that reduces the amount of cholesterol our body makes.

The soil is an excellent place to look for new molecules for two reasons. First, the soil is chock-full of microbes, holding about one-quarter of the world's biodiversity—we think— and we have very little idea what those microbes are or what they can do. For example, a research trip to the Amazon in 2020 found up to four hundred types of fungi in a teaspoon of soil. About forty per cent of them were previously unseen, not yet investigated or unnamed. Scientists estimate we only know about two per cent of what's shuffling about under our feet—and even that is a moving target. The more we discover, the more we realise we don't know.

Second, soil microbes are in constant combat with each other.

'The microbes on Earth today are the beneficiaries of a very long arms race, one that had unlimited budget and no ethics approval, and didn't mind wiping out complete species. It has been a long, ugly fight...and we can benefit enormously from it,' says Rob.

If you want to find a microbe with a penchant for killing other microbes, what better place to look than a two-billion-year-old battlefield?

'What we look for,' says Rob, when I ask how he decides which of the microbes on his agar plates might be useful, 'is microbes beating up on each other. All microbes, including fungi, produce defensive chemistry to protect themselves from competitors.'

He shows me photos of a Petri dish spotted with white and grey blobs. The white blob is peeling back at the edges, meaning the grey blob is secreting something the white blob doesn't like.

'We want to know what the grey blob is producing,' says Rob.

Microbes at war was what bacteriologist Alex Fleming found in 1928 when he discovered penicillin. A quiet and by many accounts a rather careless laboratory technician (though an excellent shot and with interesting art ideas*), Fleming returned from a holiday in Scotland to find he had left out a Petri dish of *Staphylococcus* and something else had colonised the agar. The dots of *Staphylococcus* grew everywhere across the Petri dish, except where there was an unknown blob of mould growing. The conclusion—the blob was producing something that the *Staphylococcus* didn't like. The blob was eventually identified as *Penicillium notatum* and thus began the journey to modern-day penicillin.†

---

* Fleming enjoyed painting with microbes. The images he made, such as a mother and child, people in a boxing ring and a soldier, are quite spectacular. To achieve this, Fleming needed to find microbes that made the colours he wanted, then inoculate them at different times so they matured at the same time to make the images.

† Then followed much work by scientists at Oxford University to work out what the mould juice was effective against. The original penicillin treatments required culturing 2000 litres of mould to get enough mould juice to treat a single case. Aware that *Penicillium notatum* would never yield enough penicillin to treat people reliably, scientists searched for a more productive species. In time synthetic penicillin was developed, and now you can pop a pill with almost no side effects and treat an infection that eighty years ago would probably have killed you.

This 'beating up' investigative strategy is also what started the journey to a potential replacement for morphine. About twenty years ago, someone collected a sample of marine mud from a boat ramp in Tasmania and stored it at Microbial Screening Technologies in Sydney. While studying this microbial library, Rob noticed one of the marine fungi pummelling other microbes and secreting a compound only four amino acids long. It looked an awful lot like endomorphine, which is a type of morphine your body creates as pain relief.

We don't know why the fungus makes this compound. Microbes don't have opioid receptors* so it's more likely to be a defence or attack mechanism than for pain relief. But it got Rob thinking that maybe it could be repurposed to have pain-relief uses in humans. Rob's team couldn't attract a research grant so the work has been carried out slowly in the background, with Rob pestering lots of colleagues to help out. His pestering may very well pay off. The compound, called bilaids, has passed the animal model studies in Rob's lab and has moved onto the pharmacology department for the next steps. While it is still early days, the compound shows great promise for use as a morphine equivalent—without the problems of respiratory suppression and addictiveness.

What were the odds that Fleming's absent-mindedness would lead to penicillin growth on his agar plate? What were the

---

* Opioid receptors are part of nerve cells. When opioids such as morphine attach to the receptors, they block the pain messages sent to the brain, giving pain relief.

odds that a sample of marine mud collected years ago would hold a fungus that produced a compound that Rob and his team would find so potentially fruitful? What if it had been other scientists? What if that mud had never been collected?

History is full of serendipitous discoveries. The artificial sweetener saccharin was found when a Russian chemist studying coal tar neglected to wash his hands before eating, and found everything tasted sweet. He lived to tell the tale and cash in on the discovery, even though he had to taste various chemicals in the lab to work out which one was causing the sweetness.

In the late 1950s biologist Dr Hans Peter Frey and his wife went on a holiday to Norway, exploring the Hardangervidda National Park—an alpine plateau so high that trees can't grow (also a shooting location for parts of *The Empire Strikes Back*). Frey worked for the pharmaceutical company Sandoz, which encouraged employees to collect soil samples when on holiday precisely on the off-chance that they'd contain new microbes of pharmaceutical (and fiscal) benefit. Being a dutiful employee, Frey brought back fifty soil samples, mostly from very pretty locations where the couple wanted to take photos. One of those samples held the fungus *Tolypocladium inflatum*, which produced a compound that was eventually used to make cyclosporine, now used to help counter rejection in transplant patients.

Maybe these happy accidents aren't as unlikely as they appear. Considering that the world is crawling with microbes we don't know, it might be harder *not* to find something interesting or useful. Perhaps the trouble isn't finding the microbes, but finding out what they are useful *for*. And with an unexplored microbial world, where do you start digging?

'My backyard,' Rob Capon says simply, and I'm thankful the jumpy conference call hides my initial disappointment. I was hoping for some daring adventures in darkest Peru— howler monkeys trying to steal the shovels while Rob dodged blimp-sized mosquitoes and caiman snapped at him from the water's edge. But suburban Queensland is where he looks because he doesn't need to go much further.*

'I once caught a wasp,' Rob says by way of example, 'and was going to take it outside, but decided to culture it in the lab instead.'

He found so many microbes on the wasp that he went back to the wasp nest, scraped a section of mud into a plastic bag and took it to the lab to culture. One of the microbes that had been in there, living an unobtrusive life, was producing a molecule that can make resistant cells re-respond to medicine.†

Some researchers look to history and legend for clues on potentially useful soil microbes. 'Healing soil' from Sacred Heart Church cemetery in County Fermanagh in Northern

---

* And there are other, ethics-based reasons you should not go bioprospecting in other countries, pinching microbes from their soil and making money off their natural resources.
† Cells that are resistant to certain drugs can pump out the drug faster than non-resistant cells, reducing the amount of time the drug is in the cells and meaning less chance of the drug working. This molecule slows down how quickly cells can pump out drugs, making them more responsive to the medication.

Ireland has a reputation for healing a variety of minor ailments. The reputation stems from a priest called James McGirr, said to be a faith healer, who reportedly declared before he died in 1815 that the clay he would be buried under would cure anything he'd been able to cure while he was alive. The tale persisted, and now visitors regularly collect a sample of soil to take home. Today the site has attracted so much attention that the church has put up signs asking visitors to only take one teaspoon of soil and to return it to the plot within four days. Not doing so is supposed to bring bad luck. (There is also a helpful sign suggesting prayers to be said while using the 'blessed clay'.)

Research has revealed there is some truth behind the legend. *Streptomyces* sp. myrophorea in the soil can inhibit the growth of four drug-resistant pathogens known to cause healthcare-associated infections, namely vancomycin-resistant *Enterococcus faecium* (VRE), methicillin-resistant *Staphylococcus aureus* (MRSA) (which can cause a range of nasty symptoms from boils through to lung and blood infections), *Klebsiella pneumoniae* (pneumonia with a mortality rate just below fifty per cent) and carbapenem-resistant *Acinetobacter baumannii*.

The Great Plate Count Anomaly

Clearly, huge potential lies in the bounty beneath our feet. But there is still a glaring issue facing scientists searching the soil—the Great Plate Count Anomaly. There is a saying that soil microbes don't like agar as much as microbiologists do. Agar is the gelatinous stuff that scientists put in the bottom of the Petri dish as a food source, enjoyed by microbes and by vegetarians (it is a gelatine substitute). In the lab, scientists mix agar with other ingredients to try and get microbes to grow. For example, agar enriched with sheep's blood works well for most bacteria.

Estimates are that only about one per cent of soil microbes can be cultured on agar, leaving a whopping ninety-nine per cent of potential lying uncultured there in the ground. Because of this, some scientists argue that smearing samples across agar won't lead to new antibiotics and that the readily culturable microbes have been exhausted.

I am sceptical; Rob Capon flat-out disagrees.

'Yes, if you do genomic profiling to see how many different species are in a sample it's much more than come up on the agar plate,' says Rob. 'But that doesn't mean the ones [microbes] that do come up are not useful. You can spend years trying to convince a microbe that doesn't like agar to grow. But that's years on one microbe. It's faster and there's more opportunity with the low-hanging fruit. I do what is feasible, not necessarily idealistic,' he says.

Even using agar, his team can't keep up with all the microbes they're finding. A case in point—Rob collected a termite nest from near his house over ten years ago, and the lab is still working on the microbes it yielded. Although they use traditional culturing media, they use thirty-three different types, with variations in concentrations of nitrogen and carbon. Some are solid while others are broths. Some of the broths work best shaken, others give different results if left static. The variety provides more opportunity for finicky microbes to grow.

Other companies that believe traditional culturing methods have outlived their usefulness are developing other ways to culture microbes. One company in particular, NovoBiotic, has had some success by tricking the microbes into thinking they are roaming free in the soil. Their isolation chip 'iChip' still uses agar. Soil samples are put into the small device with a semi-permeable membrane, then the device is buried in the soil. Microbes in the sample have the benefits of agar and a familiar soil environment but can't escape into the surrounding soil. It's somewhat like being in a comfortable cage with a buffet. Using this method in a grassy field in Maine, NovoBiotic has cultured over ten thousand microbes including the previously unnamed *Eleftheria terrae*, which produces teixobactin, an antibiotic that works (in laboratory studies) against methicillin-resistant *Staphylococcus aureus* (MRSA), vancomycin-resistant enterococci (VRE) and *Mycobacterium tuberculosis*.

The scientists believe that growth factors from other microbes in the soil cross the membranes and help keep the trapped microbes alive. This method breeds more domesticated microbes that are easier to move into a Petri dish.

I believe that if we are going to continue in this game of medicine discovery, both approaches help. Rob's traditional method has found a possible alternative to morphine, while the new iChip may have found a powerful new antibiotic. When there is so much that is unknown, and we're all after the same thing, it makes sense to use multiple approaches.

—

If all this talk of microbial discovery sounds like your cup of agar, there are citizen science projects around the world to speed up the process of cataloguing microbial diversity.

In Australia, Rob has started Soils for Science (https://imb. uq.edu.au/soilsforscience), which encourages residents to send in samples of soil for the lab to culture. This program was inspired by the Citizen Science Soil Collection Program by the University of Oklahoma (www.whatsinyourbackyard.org).

In return for digging up some soil from your backyard, you will be able to track on their website what they've found in the sample and see high-resolution photos of the microbes at war. And your backyard might just end up hosting the next wonder drug.

The dirty paradox

A few years ago, I helped a childcare centre set up some gardens for the children to enjoy. We were planting seedlings—snow peas, celery and winter herbs. Most of the children were about three years old, happy to pull out weeds but lacking the fine motor skills and patience to plant seedlings. One little girl, insistent on helping, cheerily smashed the celery seedlings into the soil while I tried to straighten them up again, then realised to her horror that she was 'dirty'—there was a smear of soil on her sleeve—and tears started to well. I assured her we could simply wash the soil off but her bottom lip trembled all the way to the bathroom.

Was this a learnt response? I wouldn't be surprised—I have seen many people, parents and non-parents, grimace when they find soil on their hands, clothes or shoes, then whip out the antibacterial wipes to banish the dirt.

Soil has a curious place in our lives when it comes to health. We wash it off as soon as we notice, then pay someone to slather it across our face and body. We want the medicines it yields but not the mess. Photos of hands cupped with soil to make a heart shape imply wholesomeness and life. Yet it harbours all sorts of nasties like tetanus, anthrax, heavy metals and that worm that burrows in through the soles of your feet.

The soil also holds, as researchers are finding, the key to a good immune system.

Allergies and immune diseases are some of the fastest-growing conditions in Australia. As of 2014 almost twenty per cent of the Australian population had an allergic condition.

Is it our diet? Lifestyle? Cleanliness, more caesareans, more antibiotics and less vitamin D? Air pollution and urban contaminants were blamed for a while.

Scientists believe all these factors play a part, but increasing urbanisation is at the centre of it. Fifty-six per cent of the global population lives in cities, a figure predicted to rise to nearly seventy per cent by 2050. In Australia, just over eighty-five per cent of the population lives in urban areas. As we've paved and concreted over the soil, we've also sealed off an important defence against allergies: the more we learn about immunity and microbes, the more certain it seems that the association between urbanisation and lower immunity stems less from the presence of pollution and more from the absence of nature.

Immune systems need data. Exposure to a broad range of microbes is critical to give the immune system data and to stop it from overreacting when it encounters new microbes. Microbiologist Graham Rook famously described a baby's immune system as being like a computer with 'hardware and software, but no data to process'. Rook's 'old friends' hypothesis suggests that humans evolved alongside certain symbiotic microbes, and that exposure to them is critical during infancy to develop a well-regulated immune system. Without enough exposure, the immune system gets confused and starts attacking the person themself—causing autoimmune diseases—and harmless molecules, leading to allergies.

The more data the immune system receives, the better prepared it is for pathogens and disease, or even basic allergens like peanuts, eggs, mites and so on. Unfortunately, now-outdated advice to avoid foods like peanuts and eggs during pregnancy and for the first year of the baby's life has probably resulted in lots of unnecessary allergies. There is ample evidence from Europe showing that children raised with farm animals are less prone to atopic diseases such as asthma. Even living in a town surrounded by forests and farms can reduce allergies.

There is a great study that compares allergies across the Finnish-Russian border. After World War II the Finns modernised while the Russians retained a more agrarian lifestyle. When researchers studied the health of children from each side of the border, atopic diseases like asthma, hay fever and eczema were three to ten times more common on the Finnish side. Across the line in Russia, hay fever and peanut sensitisation were pretty much non-existent.

Looking closely, the Russian children had a greater number of and more diverse populations of *Acinetobacter*, a genus of bacteria found widely in soil and water. In the soil, they live happily in the dry or the damp, and stay busy by contributing to nutrient cycling in ecosystems. On people they also seem to live happily (*Acinetobacter baumannii* accounts for most *Acinetobacter* infections in humans; it's happy, but the people with the infection are not). Exposure to soil, water and plants in day-to-day life gave the Russian children an advantage. Exposure to nature, to a rural lifestyle where you're more likely to eat a bit of soil still on the veggies and get up close with farm animals, and probably don't carry sanitiser with you, breeds children less prone to these issues.

Replicating the situation with mice gave a similar result. Researchers settled one batch of mice on clean bedding; another batch had soil sprinkled on their bedding before their cage was placed in a barn with other animals. Six weeks later the impacts on allergic responses were notable. The clean mice were more susceptible to lung inflammation from an asthma-triggering allergen while the soil/barn mice had more anti-inflammatory proteins. The soil mice also had an altered microbial signature in their guts. More Firmicutes and fewer Bacteroidetes are usually associated with asthma and inflammation but these rugged barn-raised mice had flipped the populations to have fewer Firmicutes and more Bacteroidetes.

It seems clear: the array of organisms in the soil and water and on animals helps train the immune system to better respond to allergens.

We have known for a while that microbial diversity is important for health. Patches of atopic dermatitis have a different microbiome to healthy skin, with more *S. aureus* and less microbial diversity than other healthy skin. As dermatitis improves, *S. aureus* decreases and the microbiome begins to reflect that of healthy skin.

We also know that microbial health and diversity in the gut is important. Once the supplement companies cottoned on, hundreds of probiotics and prebiotics started to claim shelf space in the chemist. Most of them don't do much but the idea is right: happy gut microbes, happy you.

What we are now learning is that there is a soil–gut connection, and part of our microbial gut flora comes directly from the soil. In one study, researchers placed a tray of soil outside a mouse cage and ran a fan to blow gently over the soil and towards the cage. After seven weeks the soil mice were less stressed than their clean-living comrades and their gut microbiomes were more like the soil. The soil microbes had wafted their way across the air currents into their little mouse faces and colonised their guts.

The biodiversity hypothesis states that contact with natural environments enriches the human microbiome, promotes immune balance and protects from allergy and inflammatory disorders. Research on the human microbiome really only started to ramp up in 2007, with the Human Microbiome Project, yet we have learnt a lot in those years, the main take-aways being:

1. Every human is basically a walking host for microbes
2. Keeping those microbes diverse and happy is important for our health
3. Try as we may, no amount of pills or potions or lotions will replicate what we get from natural exposure.

Which raises a two-pronged problem. Our urbanisation and busy indoor lives mean we spend far less time in nature than we evolved to. Once upon a time we would have met our old microbial friends going about our daily lives: foraging, growing food, not quite washing all the soil off—generally being more connected to nature.

Second, some scientists believe soil biodiversity is dwindling—so even if we do get a healthy dose of nature, said nature isn't as good for us as it once was.

There is a growing idea that biodiversity loss leads to immune dysfunction and disease. Certainly, less exposure to nature leads to lower gut diversity. You can see the trend in animals where domesticated and zoo animals have lower gut microbial diversity than their wild counterparts. So what does it mean if biodiversity, especially soil biodiversity, is dwindling? Will our health continue to degrade at the same rate the soil degrades? Will more diseases increase as wilderness decreases?

We will all just have to wait and see what research and time uncover. We also don't know which of our old microbial friends does what, if one is more beneficial than another, or whether, as is more likely, better health stems from regular exposure to a wide variety of wildernesses.*

So where does all that leave us?

Given that we are not all returning to an agrarian lifestyle any time soon, the next best thing is to bring these practices into our cities and urban areas: we want greener cities with more public open space, gardens, backyard veggie patches and as many 'nature' spaces as possible. We also need to preserve the nature spaces we have left, and thoroughly enjoy them.

Play sport outside; sit under the shade of a tree; maybe start a small garden—herbs if you like, or flowers to attract pollinators. Or just grow something you think looks nice. Visit farms if you can, milk a cow, pat a goat and walk in the forest. Get grubby and wash yourself off afterwards. Then do it again. Sink your toes into soft mud, build a sandcastle, try pottery, go on a nature scavenger hunt. Go birdwatching, get a worm farm. Simply lie in the grass and while you're lying there, the grass tickling your heels and ankles, feel the thin layer of soil beneath you, supporting everything above.

---

* Either way, do not eat soil as a health remedy. You can't tell what is in any particular soil without copious tests. It harbours old enemies as well as old friends, and eating those enemies isn't a way to deal with them.

Geophagy

WHEN I WAS eight, my cousin and I convinced my five-year-old brother to take part in a sand-eating competition. We sat in the backyard cubby house with red plastic plates filled from the sandpit. On 'go', my brother started shovelling spoonfuls of sand into his mouth, wincing but persisting, while I tipped my plate of sand on the floor under my chair and smugly told him I had won. (Which I had.)

Eating soil is not only a habit of children, whether by choice or as the dupe of an evil sibling. Eating soil, known as geophagy—from the Greek *geo* (earth) and *phagein* (to eat)—is a practice that transcends time and culture. There is evidence from every continent and most cultures at some point in time that people ate soil. These days it is most common in sub-Saharan Africa, parts of South America and the southern USA.

In Western culture geophagy is considered an eating disorder, falling under the umbrella of pica: the desire to eat non-food items such chalk or clay.* But depending on where you live, eating soil is a daring foray into new gastronomic heights or a normal part of life. For those wealthy consumers willing to spend at least fifty bucks on a cup of coffee brewed from beans shat out by a small mammal,[†] soil is positively tame. A few restaurants have tried to infuse dishes with soil—as a soil vinaigrette, or eggs cooked in soil for a 'campfire effect', but the trend hasn't really taken off. Indeed, those partial to eating soil seem to prefer hardened chunks rather than a sprinkle as seasoning.

There is no one agreed-upon reason why people eat soil but there are multiple theories. Some believe it stems from hunger, although plenty of well-fed people indulge in soil. We do know that it is most common in women of childbearing age, and particularly pregnant women, who use it to alleviate nausea. How it works is unknown, possibly because we also don't know exactly what causes nausea and vomiting in pregnancy. (And speaking as someone who had a horrible fifteen weeks with twenty-four-hour pregnancy nausea: when you eventually find something that provides relief you don't care how it works, just that it does.)

There may be a health angle. For centuries, clay was pressed into tablets and sold all over Europe as an antidote to

---

* The name pica is the Latin word for the Eurasian magpie, a bird once thought to eat a variety of shiny non-food objects like bottle caps.
[†] Kopi luwak, or civet coffee, is made from coffee berries that have been eaten and excreted by the Asian palm civet.

poisoning. Called *terra sigillata* ('clay bearing little images', though usually roughly translated as 'sealed earth'), these tablets could fix everything from stomach ulcers to snake-bites. Much of the clay in Europe came from the Greek islands of Lemnos (Lemnian clay) and Samos. By the 1600s in Paris, Lemnian clay was as popular as mummy flesh (sold as a remedy for internal bleeding)—and easier to come by.

That clay could counteract poison is not a completely ridiculous idea. When researchers poisoned rats and offered them food or clay or kaolin, they always chose clay or kaolin. The more poison the rats were given, the more clay or kaolin they ate. As rats can't vomit, it was speculated that this was their way of trying to soak up the poison. Clay particles are covered in negatively charged sites that attract positively charged ions. So in theory clay particles could catch and hold on to a poison, but being able to reliably use it to counteract poisoning is a theory fraught with assumptions. The poison has to bind to the clay before it does too much damage; it has to be a poison that only works after it gets into your stomach; you eat the clay in time; the clay can adsorb enough poison to prevent death or serious illness; the clay itself doesn't contain pathogens...and so on. We can't experiment on people to find out and I don't recommend using yourself as a test subject.

There is, however, some interesting research from the 1990s on birds, in which clays were shown to help counteract the intake of alkaloids.

Many birds eat soil for grit. They have no teeth so they use coarse soil or small pebbles to help grind food in their gizzards. Generally, the bigger the bird, the bigger the particle size they prefer: half a millimetre for sparrows, 3.5 millimetres for crows and so on. Researchers noticed Peruvian parrots favouring specific soil horizons along the banks of the Manú River in south-eastern Peru. (The paper doesn't say which horizons, but it wasn't the soil at the top or the bottom of the soil profile; it was somewhere in the middle.) When tested, the soil was about fifty per cent clay with very little sand, so the particles were too small to be any use as grit. The soil was less nutritious than the food the parrots ate, so the parrots weren't trying to supplement their diet. What the soil did have was a high cation exchange capacity (like Fuller's earth, mentioned in the Health chapter) and an ability to adsorb compounds with a low molecular weight. The researchers proposed that the parrots were eating this soil to counteract the high levels of alkaloids in the seeds and unripe fruit they ate.

Many plants produce tasty, bright-coloured fruit as a lure to attract animals that will help spread their seeds. Infusing the seeds with chemical defences like alkaloids that make them bitter and toxic is a good way for animals to learn to eat the fruit and discard the seeds, giving the plant a chance to reproduce. The unripe fruits and seeds in this case contained quinine (the ingredient in tonic water formerly used to prevent malaria), strychnine and tannic acids: toxins that, if you or I ate them, would probably spell our demise. Being able to eat fruits that few other animals would touch gives the parrots an advantage when there is competition for food. And they can eat the fruit because they also eat clay. Blood tests on the parrots showed that birds who received clay with an oral dose of quinidine* had sixty per cent less quinidine in their bloodstream after three hours than birds who received only water with their quinidine. When researchers exposed brine shrimp to seed extracts, most of the shrimp died. Adding clay yielded a three-fold reduction in toxicity.

These results are fascinating, but I would caution against using experiments on parrots and shrimp as evidence that you should eat soil as a health remedy. Apart from the usual issues with applying animal studies directly to humans, clay can contain and release 'toxins' such as heavy metals, as well as bind them. They can also bind up useful metals like iron, magnesium and zinc. Which leads us to another possible reason for geophagy: micronutrient deficiency.

The link between anaemia and eating soil has been known for centuries. In sixteenth-century Europe, geophagy was considered a symptom of a disease called chlorosis, the 'green disease', which sometimes imparted a green tinge to the skin and mainly affected young teenage girls.† It was likely microcytic hypochromic anaemia: a form of iron-deficient anaemia where the red blood cells are more pale than normal because there is less haemoglobin. Haemoglobin molecules have iron in the middle.

But which came first, anaemia or soil eating? Does the

---

* The researchers used quinidine instead of quinine because it is less toxic to animals and the assay test methods are more sensitive.
† A recent case of a young girl with a green tinge was treated with iron salt therapy. The treating physicians state that we still don't know what causes the green colour as it is not always present, even in extreme anaemia cases.

lack of iron trigger the desire to eat soil for some, or does eating soil cause iron deficiency?

Studies on geophagy test people who have been eating soil for a while (again, we can't experiment on people to find out). Clay both captures and releases ions, so while it does contain micronutrients, it is just as capable of binding them as supplementing them.

The strongest argument for geophagy is cultural. Where it is practised, it is largely a cultural norm. Wander through the street markets of Cameroon and you'll find eating clay for sale next to the veggies, and in some places it's perfectly acceptable to take a piece of clay from the wall of your house and eat it. You can buy eating clay (though it is not labelled as such) from grocery stores in the USA and London, and more frequently online.

The documentary *Eat White Dirt* briefly explores the practice in Georgia in the USA. Some people have a coffee and a cigarette for breakfast. Others, like Tammy, an interviewee, have a chunk of white kaolin and a can of Coke. Tammy loves kaolin, snacking on the hardened pieces like someone might snack on a bag of almonds.

More interesting in my opinion than the reason—whether cultural, disordered, hunger based or nutritional—are the process and effect. Does different soil have different flavours? Do you have to prepare it somehow? And what does it do to your body—your bowels and teeth?

The reviews from online stores are enlightening. Gummy, crumbly, crunchy, gritty, creamy. Texture seems to trump taste, though taste is still important. You can buy clay plain or seasoned, or roasted or smoked for extra flavour; salt and vinegar, black pepper, cardamom and truffle are just a few of the options. The most popular clay on one site is the rose clay, which one reviewer writes has *earthy and metallic notes and a lovely crunch. Gets very smooth in the end which is not bad at all and it does not stick. Just a little creamy which is easy to clean up.* Another consumer commented that she likes to follow up a junk binge with some clay, as it makes her feel healthier.

It's true that one characteristic of pica is a strong craving, like a craving for tobacco. Yet, while those who eat soil may experience cravings (as anyone might for something they enjoy), these are not the words of people eating only out of compulsion. They are connoisseurs. No experienced clay eater is going to just dig a hole where they stand, because not just any clay will do. A *New York Times* article from 1984 noted that in Cruger, Mississippi, the clay-rich hills are preferred to the grittier soil on the flats. From a soil-science perspective, the soil on the flat landscape is alluvial, washed in over time, and therefore likely to contain larger particles, giving it the gritty texture over the smoother clay. The alluvial soil is also likely to be younger than the older, more weather-beaten clay on the hills, where the larger gritty particles have been broken down naturally over thousands of years.

I am yet to try eating clay but I did find a video of a newbie giving it a go. He prepared the clay by microwaving to make it very dry and crunchy (baking has the same effect, and the additional benefit of killing any pathogens that may be lurking in the clay). He found the clay very dry to start, sucking the moisture from his mouth. After a while, when the clay mixes with saliva, it becomes very smooth and creamy, and has little taste. 'Inoffensive' was the verdict. Another newbie described it as forming a paste that stuck to the roof of her mouth, 'not awful but not great either'.

Overindulgence of any kind is unwise. Chew enough gum and you'll get repetitive strain injury in your jaw; even too much water can kill you. Those who eat clay say they crave it and enjoy eating it, but there can be some very unpleasant long-term effects. In extreme cases, where so much clay is consumed that nothing can move through the digestive system, you can get impaction of the gut, requiring surgery or otherwise leading to an untimely demise. Or, less drastically, eat gritty soil for long enough and it will eventually damage your teeth—a British woman lost four as a result of daily soil consumption to ease stress. Coarse grains will erode tooth enamel, but there may be a secondary issue with iron deficiency. In agriculture, high pH soils cause iron deficiency in plants, so perhaps the same thing happens if you eat alkaline soil (so maybe geophagy does come first in some cases of anaemia). The body uses iron to build teeth, hair, skin and nails. Those with iron-deficient anaemia tend to have greater hair loss and a higher risk of tooth decay.*

---

* Beavers aren't known for getting caries, an infectious disease that can lead to cavities. The yellowing of their teeth is caused by the iron in their tooth enamel, which provides even better protection than fluoride against cavities.

The more serious concern with continually eating soil is that you may also be eating contaminants. People who promote eating clay to cleanse your digestive system of toxins seem to ignore the fact that clay can also contain them. Soil is highly variable. Unless every container of mined clay is analysed you don't know what you are putting in your mouth. Eating it for a day or two might not hurt, but ongoing consumption could easily lead to accumulation of something nasty. Or not—but that's my point. You don't know unless you test. Baking or microwaving kills pathogens (hopefully) but it won't do much with heavy metals such as lead or arsenic, or other contaminants that could be present. One online clay shop claims that their clay is mined deep underground, away from contamination. This might help avoid modern contaminants like pesticides, but lead and arsenic are both found naturally in soils. Lead is in the rock the soil has weathered from; digging deeper might even mean increasing lead levels. Young children who like to explore by putting things in their mouths are particularly prone to toxic effects from lead because they ingest it on the soil particles and because they absorb four to five times as much lead as an adult.

As with any other habit, people give up eating clay. Often they don't want to, but do so out of concerns about social stigma or health issues. One woman said her husband complained it made her mouth taste like mud. Tammy from the documentary found her constantly white lips and mouth embarrassing, but admitted her habit was incredibly difficult to break.

I doubt I'll be adding clay to my diet, but I do plan on trying a piece. If it seems like an odd practice, that's just a question of cultural norms. You might ask yourself: is it really that different from chewing gum?

Beauty

THE MOMENT I realise I'd make a terrible spy I'm lying on a table, stripped down to my knickers, and struggling to come up with a cover story (despite having had weeks to do so) when I'm asked why I'm so interested in clay.

I'm undercover at a fancy spa retreat in the Byron Bay hinterland, booked in for an exorbitantly priced 'body masque'. Soil, specifically clay, has been used for thousands of years as an aid to beauty so, in the spirit of primary research, I want to see just how beautiful I could become if I marinated in mud for an hour.

'Red is for sore muscles, green for nourishing skin, yellow for inflammation,' my therapist, a sensible-looking middle-aged woman, says. While I ponder which oxidation state of iron to choose, she turns up the volume on Enya and lights a scented candle in the corner of the room. 'And you'll probably want to take your knickers off.'

'How do they'—the different types of clays—'do different things?' I ask.

'It's something to do with the chemistry and how clays absorb and hold on to different chemicals,' she says, admitting this is the extent of her knowledge. Then she hints again that it might go better with my knickers off, while I mutter something about us going out for a drink first.*

She changes the subject: why am I interested in clay?

It's an honest enough question, and easily predicted. 'I… uhhh…do a lot of gardening.'

She tilts her head.

'I'll take yellow,' I say meekly, and lie back down.

The therapist steps into an adjacent room and returns stirring a mixture of yellow clay powder, water and moisturiser in a ceramic pot. It looks a bit like a soft-serve ice-cream, but a dusty yellow—Pantone colour Mimosa (or, for the soil scientists, 2.5Y 8/8).

'The clay will dry and shrink, pulling the toxins out of your body,' she says as she spreads it quickly over my limbs. The clay goes on hot and cools quickly. I let her dubious comment about toxins slide—I've worked for years with soil, mostly gloveless. I should be among the most detoxed people on the planet. Unless it works both ways, and the skin's so

---

* Joke was on me, though. I resolutely kept them on, which I regretted two hours later when they were so covered in yellow clay I left the resort carrying them in my hand.

porous it lets toxins in? Anyway, by now I'm cocooned in a cotton cloth and she's massaging my scalp. She might not know much about clay, but she knows how to give a massage.

Two hours later I leave the retreat relaxed, full of questions and empty of wallet.

I get home and seek my husband's opinion. Have I emerged a beautiful butterfly?

'Your face is glowing!' he assures me.

'They didn't do anything to my face. It was a body treatment.'

'Oh.' He leans forward and sniffs my forearm. 'You smell nice?'

'That's the fancy soap they use to wash off the clay.'

Out of compliments, he shrugs and returns to his book.

I inspect my arms and legs. Less inflamed? I have no idea. Is my skin softer? Hard to tell. Maybe it had no effect; maybe I needed to use the green clay.

In the beauty department, part of the allure of clay is the idea that it can 'draw out toxins'. It's not entirely fanciful—it stems from how clay works. But like everything else in this world, reality is more complicated.

Clay particles have negatively charged sites around the edge that attract and hold on to positively charged cations, such as calcium (Ca^{2+}), magnesium (Mg^{2+}) and sodium (Na^+). This is called the cation exchange capacity or CEC (mentioned earlier, in the Health chapter). The more charged sites there are, the higher the CEC, and the more cations the particle can attract and hold. This concept is also why clay soils are more fertile than sandy soils. Sand has a low CEC, meaning it cannot hold on to as many nutrients and is less fertile than a clayey soil.

This ability of clay to attract and hold on to positively charged ions is one part of the detoxing idea. Many pollutants and 'toxins' have a positive electrical charge and can adsorb to the edge of the clay particle. What is more important is the *type* of clay mineral. Clays are not created equal: different clay minerals have varying CECs, which is partly why they are used in different products.

Kaolin, one of the better-known clay minerals (you might know it by its colloquial name China clay, a white clay used to manufacture porcelain, or...china), has quite a low CEC, as low as 5–15 centimoles of positive charge per kilogram (cmol⁺/kg), and is usually considered a 'mild clay' in terms of body treatments. It does not have much 'pulling power'. Montmorillonite clays, named after Montmorillon in France and with a CEC of 60–100 cmol⁺/kg, can have as many as seventy-five different trace elements adsorbed to their surfaces.

Different clay minerals also attract water and swell to varying degrees. Clay particles are made of layers. When there's water, some clays swell, stretching open like a sponge, and drawing water and other things between the layers (into the interlayer space). Ions that go in here are locked away more tightly than those held on the outside of a particle.

Kaolin absorbs water but with limited swelling, making it useful in diarrhoea medication, as a preservative and as a drying agent for nappy rashes. Smectites, on the other hand, have no qualms about being bloated: they love it. Bentonite (a more common name) is usually predominantly smectite. It can adsorb the odours from animal faeces (in cat litter), and is an alternative to the silica gel packets used as a desiccant to stop moisture ruining your products. It's also used by our friends in the wine industry where, because it can adsorb protein molecules from liquids, it's added to white wines to remove excess protein that would otherwise cause a cloudy appearance as the wine warms. Montmorillonite can swell to several times its original volume. It's good to plug holes underwater—in dams, for example—but dangerous in tunnels; and it can cause landslides.

Trees are often blamed for causing cracks and movement in buildings, but the culprit could be a swelling clay or a combination of both. Around Moree in northern New South Wales there are some clays that, when dry, have cracks many centimetres wide—wide enough to lose a small mammal down—but with enough water the clay swells so much the cracks disappear completely.

The undoubted ability of clays to attract and hold ions and water is probably what started the idea of the toxin-catching superpower. My question: how does clay know what constitutes a toxin on human skin?

It doesn't. Clay doesn't know or care what a toxin is to you.

What a clay adsorbs depends, among other things, on what mineral it is and what ions are around it in the soil. For example, kaolinite prefers lead to calcium, but montmorillonite might be the opposite. Their affinity for, say, metals can vary with what metals are present in the soil. For example, illite is more effective than montmorillonite at adsorbing copper when there is lots of copper around, but montmorillonite is better when copper concentrations are low.

Which raises that question again: if clay isn't too set in its tastes, will it also draw all the 'good nutrients' out of me? It's a concerning thought. After all my time working with soil, I don't have nice hands. Have there also been some unwanted nutritional side effects?

Getting an answer any more detailed than a marketing claim takes effort. One beautician tells me that the positively charged good minerals in the clay will swap with the positively charged toxins on my skin. (Not buying it.) Another says that the clay takes away the toxins and then you add nutrients afterwards with moisturiser. (More feasible.) I eventually get some answers from cosmetic chemists: the scientists who formulate cosmetic products.

In short? No. Clay applied to your skin won't affect your insides.

Clays work on the epidermis, the top layer of the skin. One of the skin's main functions is to act as a barrier to protect bad things from getting in and to prevent water from getting out. It's actually quite challenging to deliver something through the skin barrier because the top layers of your skin are so densely packed.

As the clay dries on your skin it removes some oils and

might adsorb some unwanted particles sitting on the surface. Peeling or washing it off takes some dead skin cells with it, so there's a mild exfoliating effect. As for adding the 'good' things, the answer lies in what active ingredient you are trying to get into the skin. Vitamins A and C, for example, work on a concentration gradient so they aren't as affected by the charge on the clay. It's a bit like a fridge: magnets (cations) will stick to the fridge but the saucepan (the active ingredient) won't.

What about the different colours? For my body masque I was offered red for sore muscles, green for nourishing skin, and yellow for inflammation. Soil colour mostly comes from various oxidation states of iron. Do these different coloured clays really have different effects?

In this instance, no. As we've seen, colour is important in the antibacterial uses of clay, but in the cosmetics arena colours don't have much effect. Red, yellow, grey and blue-ish grey all stem from iron. In very low-oxygen environments, iron forms in a reduced state and takes on drab green and grey colours. When there is more oxygen available the colours become more yellow, then red, and brighter: the iron is effectively rusting. The colour isn't related to the clay mineral or the CEC. You can have red kaolin, for example.

In any case, most cosmetic formulas have inorganic pigments added to them to enhance the colour. Cincinnati-based cosmetic chemist Kelly Dobos tells me that titanium dioxide lightens and brightens, while chromium greens or ultramarines are used to create colours that are more appealing to consumers, and reinforce marketing concepts. Kelly assures me that colour is a huge part of cosmetic formulation. Consumers believe a clay's colour is important and want to attach meaning to it; marketers know and use this.

Colour helps with what Kelly calls 'imagery ingredients': something added because it helps the consumer imagine or perceive a benefit. One example is the addition of green tea to face masks. 'That sounds nice on the front of the label,' says Kelly. 'But if you flip over the package and look at the list of ingredients you see that it is green tea extract, and it's near the preservative.' In other words, it's a very low dose.

Consumers know green tea has antioxidants when you drink it, but when it's used on the skin as an extract, and in such a small amount? 'It's really not delivering any benefit to the consumer. It's basically fairy dust.'

She goes on, 'I've seen a mask that claimed to contain amethyst as an ingredient, and it was purple. They used lots of ultramarines to make it look purple.'

Colour is only one aspect of the 'theatre' of beauty products.

'When we formulate cosmetics there are a lot of sensorial cues that consumers will notice and attribute to some sort of efficacy,' says Kelly. It might be the colour, the texture or the sensation on the skin. In the case of clay masks, as the clay dries it feels like it is tightening the skin. In a lot of clay masks you can see small spots of oil around pores—another visual cue that allows marketers to say things like 'removing toxins'.

'But formulators know it's not really pulling toxins out of the skin. It's absorbing some of that sebum,' says Kelly. In at least one case, a company added a menthol ingredient to a clay mask so it would give a tingling sensation on the skin. 'That really resonates with consumers,' says Kelly. 'It makes them feel like the product is working.'

Although there is a lot of 'marketing puffery', as Kelly puts it, in the cosmetic industry, this does not mean clay is a useless ingredient. Clay is common in cosmetics for a reason: it has properties that make it very useful. It can bind ingredients together, can adsorb oil and water, and can help improve product consistency. Bentonite clays are useful as cleaners (to mop up oils on the skin), or as a gum to help thicken and stabilise a cream. Clay is a rheology modifier, able to change the texture and viscosity of certain products: to make them flow, pump or suspend particles, for example.

Suspending pigments is particularly important in the nail-polish industry—you want the polish to spread easily when brushed. In the 1940s and '50s, Kelly tells me, before formulators worked this out, nail polish was made using technologies from car paint. The particles would settle in the bottom of the bottle, so manufacturers would try to hide it with the packaging: putting the label over it or using different-shaped bottles to minimise the look of the pigment settling, until they discovered they could use clays to suspend the particles. For toothpaste, clay helps it sit in a nice shape in the tube, but also squeeze easily out of the top.

—

Most people feel better after they take a trip to the spa or have a facial, but it's not really about the clay or mud. The benefits more likely stem from downtime and mental relaxation than from any powerful clays. But even though the effects might be indirect, they are still worthwhile. Stress kills. Sometimes one of the best things you can do for your health is stop, relax and rest.

If a hot mud bath helps you do that, and you can handle the price, go for it—but know that it's the heat that is likely doing the bulk of the work. There is plenty of research on the benefits of saunas, including on mental and cardiovascular health, and of heat packs for arthritis and rheumatism. Dr Hunnius, who famously started the spa culture in Haapsalu, Estonia, used to visit fishermen's houses and noticed some kept their feet in warm mud to relieve pain from arthritis. Was it the mud, or the heat? Sand baths in Egypt and Morocco, where people are buried up to the neck in sand, have been used for thousands of years to help with arthritis. The sand itself isn't particularly special; its capacity to stay hot is.

The point is that a placebo effect is, as they say, still an effect. Beauty treatments exist to make you feel good, and it doesn't much matter how they get there. If the odd clay treatment does some mild exfoliation or soothes a stressed soul and not much more—so what?*

---

* And to the clay marketers out there: if you are going to use the word 'harvest' when you talk about mining clay, it needs to be in the sense of harvesting organs, not wheat.

Making soil 2: extra-terrestrial adventures

THINK BACK TO the scenario at the beginning of the book, where you were lost in a vast wilderness. It might have been a struggle, but you found everything you needed to survive.

Now imagine you are on Mars. Apart from the obvious threats to life—no oxygen, no atmospheric protection from deadly radiation, little gravity and night temperatures averaging minus sixty degrees Celsius—survival based solely on local resources just got a whole lot harder. There is no food to forage or hunt, no reeds for your basket or heather to stuff your pillows. You can't start a fire—there's no kindling unless you brought it with you from Earth, which would have been pointless because fire needs oxygen. There is water if you can somehow melt some of the ice caps—but is it drinkable without treatment?

Despite the enormous challenges Mars presents, space agencies and private companies are currently working out how to set up a colony there. Whether for tourism or dooms-day scenarios, Mars is very sexy right now.

And it has a few characteristics that make it a feasible, albeit difficult, prospect. It is Earth's closest planetary neighbour and has a similar day-night cycle (a twenty-four-hour-and-thirty-nine-minute day, although a year is twice as long because the orbit is twice the size of Earth's). We won't burn to death like we would on Venus, with its day tempera-tures of 462 degrees Celsius, though we might freeze. It is the only planet in our solar system where greenhouses could make use of natural sunlight, though the light intensity would be low, similar to Alaska's. And, yes, there is water in the ice sheets in the polar ice caps. There are challenges, but to many people they don't seem insurmountable.*

With the sheer cost of transporting anything from Earth to Mars—about five thousand dollars just for a lemon—colo-nists won't be packing a lifetime supply of food, bedding and DVDs. Martian colonies will need to use, as much as possi-ble, the limited resources available to generate the five major consumables—food, energy, oxygen, water and construction materials. Humans are very clever, but even with all our space

---

* Interestingly, suggestions to terraform Mars are very similar to what we're trying *not* to do on Earth: melt the ice caps at the poles to release huge amounts of CO_2 and water into the atmosphere so the atmosphere develops and warms.

technology and ingenuity, we can't make something out of nothing.

Luckily, although Mars only has three main resources—sunlight, ice caps and rocks—we can use them to make others. Energy, probably most readily derived from sunlight and supplemented by nuclear fission, is the critical starting point from which other necessities, including oxygen and water, can be generated or manufactured. Energy would run the robots that would be sent over first. With energy we can (we hope) harvest oxygen from atmospheric CO_2*—somewhat like our clever plant friends—or from the briny water on Mars, melt the ice caps to get water, make more energy, and have pressurised living environments, lights, heat and so on.

Food and construction will rely on Mars' other major resource: rocks and the red dust that swirls around the planet in mighty storms and that gives the red planet its characteristic hue and nickname.

Where Earth's soil has five ingredients—minerals, air, water, organic matter and living things—Martian soil is almost all mineral, with about two per cent water. It is quite sandy, with layers of very fine dust, and may have some tiny pores that air and water could inhabit if they were present. As far as we know, it does not contain organic matter or life—so is it soil? Or just rock dust? In a quick poll of a soil-science group, just over half said, yes, Mars has soil. One quarter

---

* The technology is still experimental but the Perseverance rover that landed in 2021 carried an experiment called MOXIE, which is going to try to split the atmospheric CO_2 into oxygen and carbon monoxide via electrolysis.

were unsure, because Martian soil does not contain organic matter or life. Twenty-two per cent said no—no life, no soil—and in this section I am going to side with them, and call it regolith rather than soil. Martian regolith is far from a great foundation to set up a new society, but it is what future Martians will have to work with—and the research to turn this red dusty devil into something more useful (to us) is well underway.

Food

'I didn't pee on them every morning if that's what you're thinking,' Wieger says.

It was.

When I learned that Dutch scientist Wieger Wamelink had been using urine to fertilise beans grown in Martian regolith,* it was naturally the first scenario that popped into my head. Only briefly of course. I'm not sure you could run a properly controlled trial using this method (what if he was dehydrated one morning?) and I suppose Human Resources might have a few issues.

Wieger has been trying since 2013 but let's face facts:

---

* It is important to note that, so far, all Martian soil experimentation is theoretical. No one has yet brought back a sample of Martian regolith, though that may happen while this book is being printed. Scientists are working with analogues: simulant soil made on Earth by mixing different soils to mimic what we know about Mars. Some come from Saddleback Mountain, a twenty-million-year-old basaltic lava flow from a volcano in the Mojave Desert with iron-rich basalt; another source is volcanic soil from Hawai'i. Before any real efforts are made to move to Mars we will need to do these experiments with the real thing.

Martian regolith is simply rubbish for plant growth. We've just mentioned the lack of structure and organic matter. It also compacts easily, sets hard, is hydrophobic, salty and very low in nitrogen, and contains varying amounts of perchlorate, a chemical that some plants take up happily (such that phytoremediation* could be one way to remove perchlorate from soil), and which is toxic if humans eat enough of it. It could make a reasonable cricket pitch, but you might as well start trying to farm in pavement.

Plants need about sixteen elements to grow. Thirteen of these—nitrogen, phosphorus, potassium, calcium, magnesium, sodium, sulfur, copper, iron, boron, magnesium, chloride and zinc—plants get from the soil. Nitrogen is the nutrient plants use the most. All of these have been found in Martian soil, but some, especially nitrogen, in such low amounts that the only way of ensuring plants get enough nutrients to grow is to fertilise.

That's what we do on Earth to solve low nitrogen: we add fertiliser—often urea, a manufactured form of nitrogen developed in the early 1900s when Fritz Haber and Carl Bosch worked out how to squeeze it from the atmosphere at an industrial scale (our atmosphere is seventy-eight per cent nitrogen). Today, we've made so much of it that anywhere from thirty to eighty per cent of the nitrogen in our bodies was created by the Haber-Bosch process.

Like building materials, fertiliser is too heavy to transport

---

* Using plants to reduce the amount or effect of contaminants.

to Mars, especially given that, after a while, there will be sources available locally. Martians will use the age-old fertiliser sources: manure and urine. For most of human agrarian history, crops were fertilised with manure—cow, bat and human. All the food you eat and excrete contains elements plants can use to grow.

Urine is a wonderful source of nitrogen, phosphorus and potassium, but raw waste is highly variable and potentially hazardous. Wieger has been using struvite, a processed form of urine that ends up as granules. It's safe to handle and doesn't smell. The source? Portaloos from music festivals in Amsterdam. The toilets separate 'the pee from the poop', as Wieger puts it, and a processing facility turns the liquids into struvite.

Because of the source, says Wieger, 'you can imagine that there is more than just normal pee and poop in it. And you also have to be very careful with medicine, for instance, because we know that there can be remnants of medicine in urine'.

I like how he specifies medicine. 'Medicine' would be my biggest concern, too.

With this nutrient boost, all plants grew well, most more than twice as well as their non-struvite counterparts, especially the beans. As an added bonus, researchers found that the beans—even in the control plots with no struvite—had formed nodules showing rhizobia bacteria. Rhizobia bacteria are nature's way of capturing nitrogen from the atmosphere and putting it into the soil. Microbes from the *Rhizobium* genus 'fix' the nitrogen from the atmosphere and convert it into a form that plants can use. These clever little factories can generate anywhere up to four hundred and fifty kilos of nitrogen per hectare per year if the conditions are right. They tend to form symbiotic relationships with legume crops such as beans, which is why farmers tend to rotate cereals with a legume crop. They probably can't capture enough nitrogen to grow the yields Martians will need, but every little bit counts—it is possible that on Mars, in a pressurised environment with a simulated atmosphere, these rhizobia bacteria might be able to help out.

Although all the other nutrients plants need have been found in Martian regolith, the question is whether plants can use them: just because an element is in the soil, doesn't mean the plant can take it up. There are complex interactions between soil chemistry, nutrient form and soil biology that affect whether a specific element is available for the plant to absorb. For example, most plants prefer a soil pH of about 6.5. At very high or low pH certain nutrients become more available, or less so: there is more aluminium and less magnesium at low pH and more potassium but less iron at high soil pH. Soil microbes play a huge role, breaking down organic matter and releasing nutrients, converting nutrients from unavailable to available forms. For example, plants find it easier to absorb nitrogen as nitrate (NO_3^-) than as ammonium (NH_4^+), so nitrifying bacteria, which convert ammonium to nitrate, are obviously useful.

Heavy metal uptake is in a similar situation—the very big and practical difference being that there are plenty of metals in Martian regolith that we do not want plants to absorb too much of and transfer into the food we eat. Ingesting contaminated produce is one of the main ways heavy metals get into our bodies. For some people who live close to mining areas, locally grown vegetables have an unfortunate tendency to accumulate higher than desirable amounts of zinc and copper, as well as highly toxic metals such as lead and cadmium.

The risk is highest in leafy greens, particularly leafy vegetables like spinach or kale. These veggies are recommended as a good source of iron for vegetarians or vegans, but could take up too much from the iron-rich Martian soil, which would cause issues for human health. Although we need iron, our bodies are not very good at getting rid of any excess. It starts with stomach pain and nausea, maybe some vomiting, and if too much iron is consumed it starts accumulating in your organs, increasing the risks of arthritis, liver problems, diabetes and in some cases heart failure.

So rather than speculating *if* a metal could be taken up by plants, the easiest thing to do is test the soil water for metals. If they aren't in the water, the plants can't take them up. Analysing the soil solution several times, Wieger did not find heavy metals present above expected levels—a good sign that the metals in the soil are staying in the soil. However, one of his PhD students found molybdenum toxicity in peas after he lowered the pH enough for molybdenum to become soluble, go into the water and be taken up by the peas. There will be a lot to watch and balance as we start playing with soil chemistry and biology.

While there is still a lot to learn about growing plants in Martian soil, we do know that there are not enough nutrients in the soil to support long-term crop growth, so the soil will need fertilising with all the essential nutrients except iron. But the thing is, farming—regardless of which planet it is on—is a nutrient-export activity. Every harvest removes nutrients from the paddock that need to be put back somehow, whether through organic or inorganic forms of fertiliser. On Earth, we tend to mine for fertiliser. On Mars, recycled nutrients are the logical place to start.

A recipe for soil

In the 2015 movie *The Martian*, astronaut Mark Watney (played by Matt Damon) is stranded on Mars. Left behind and believed dead when his crew escape a severe dust storm, Watney seeks refuge in their temporary Mars accommodation. With the food left behind by the crew and some rationing he calculates he can survive for four hundred sols (Martian days). The problem is that even after NASA realises he is alive and starts to mount a rescue mission, it will be an extra thousand sols until it gets there. Watney needs to find ways to generate enough food and water to last until then.

Luckily, he's a botanist—a helpful start in working out how to grow food on a planet where nothing grows. As he says in the movie: 'In the face of overwhelming odds, I'm left with only one option. I'm gonna have to science the shit out of this.'

Which he does—with actual shit. Using packets of freeze-dried faeces from the other astronauts and some of this own (and stuffing his nose with earplugs to get the job done), Mark mixes the faeces with water and mixes the resulting sludge into the regolith, sprouts a few potatoes that were being saved for Thanksgiving, and begins turning the rock dust into productive soil.

1.   Add a hefty dose of organic matter

Organic matter is very important. It increases the amount of water and nutrients the soil can hold, is a store of slow-release nutrients as it breaks down, helps buffer changes in pH, and is food for soil microbes.

When Wieger first started working with Mars simulant in 2013 he did not add anything—he just put seeds straight into the simulant soil. Until then, simulants had only been used for rover testing or materials analysis. 'No one even knew if seeds would germinate.'

They did, but the roots had trouble growing into the hard soil and scavenging nutrients. The soil was also hydrophobic—water tended to pool on top rather than soak in—making it hard to get it to where it was needed.

Adding organic matter in the form of grass clippings helped.

'We added up to fifty per cent organic matter'—creating a mix of fifty per cent soil and fifty per cent clippings—'and that works very well, but even then the roots like to grow at the side of the pot instead of into the soil.'

Sources of organic matter on Mars, at least to start, will be more like those that Mark Watney used. Living on Mars, while far from 'natural', will need to adopt one of nature's key principles: there is no such thing as waste. Perhaps even the word will become an archaism as everything comes to be viewed elementally. All food, excrement, fallen leaves, bioplastics—even human bodies—are a source of valuable nutrients: a store of essential elements that cycle through and sustain life. Those elements need to be put into the Martian regolith to begin the transformation to soil.

To Wieger, 'One of the most important things you can do on Mars is recycle, recycle, recycle. All the parts you don't eat, all of your poop and pee—and maybe even you, if you die there—need to become the new nutrients for the plants.'

2. Sprinkle in microbes

Organic matter is food for soil microbes. Soil is a system, a balance of chemistry, physics and biology. It hosts the underground ecosystem necessary to break down wastes, keep plants fed, provide resilience to pests and diseases, capture nitrogen from the atmosphere, and more. The relationship between plant roots and microbes is tight: once crops are growing they release exudates from their roots that feed the microbes too and help keep the soil system working.

When researchers at Colorado State University added the nitrogen-fixing bacteria *Sinorhizobium meliloti* to Martian simulant soil, plants had up to seventy-five per cent more roots and shoots than those without added microbes. Interestingly, the amount of nitrogen in the soil didn't increase.

Which microbes get taken to Mars is up to the scientists to debate. At the moment, everything is disinfected before take-off and there are legal issues about taking microbes into space. These will need to be sorted first. Because humans are basically walking microbe factories, there will always be stowaways but it could be that growing food on Mars would be an opportunity to leave soil and plant pathogens and diseases behind on Earth.

3. Add a splash of macrofauna

After adding organic matter and microbes, bigger soil critters will get the engine room cranking. Insects, bugs, worms and thousands of other soil macrofauna do the heavy lifting of breaking down sticks, leaves and corpses. Their underground burrows help mix these nutrients into the soil and create channels for water, air and plant roots. To many gardeners, earthworms are a classic sign of a 'good soil'. These graceful creatures munch their way underground, creating channels and pore spaces for roots to grow, and leaving behind worm casts, which are a great source of plant nutrients. Having earthworms—or, as they should probably be called, 'Marsworms'—would be an important part of the regolith-to-soil conversion process.

Testing if worms can handle living in Martian regolith is another aspect of Wieger's work. So far, the worms are comfortable enough for the odd spot of procreation: he was delighted one morning to find a pot with four large worms and two small baby worms.

The tricky part has been collating clear experimental data.

'They are real escapists,' he says. 'They don't travel that far but they climb out of our pots. Every morning when I checked I found worms outside their pot and of course I couldn't put them back because I didn't know from which pot they came.'

Two layers of Velcro on the inside of the pots didn't deter the Houdini worms—Wieger had hoped it would be too uncomfortable to crawl over. Yet the experiment was deemed a success. On top of the baby worms, about one-third of the worms originally put in were successfully recounted. About one-third escaped, meaning one-third are missing presumed dead. When worms die it takes about three months for them to become part of the soil.

4. Remove excess salt

What do fireworks, explosives, rocket propellant and Martian regolith have in common? They all contain perchlorate, a salt compound consisting of one chlorine and four oxygen atoms. Perchlorate is only usually found naturally in very dry places like the Atacama Desert in Chile.

On Mars, it is abundant.

Experiments on Earth have shown some plants willingly take up perchlorate. So far it does not seem to harm the plants themselves (though some freshwater wetland plants struggle) but is toxic to those who try to eat the plants. This isn't always a big problem: fruiting plants like tomatoes usually store such 'unwanted' products in their leaves and the other parts that won't get eaten; the goal after all is to have the fruit eaten and seed dispersed to ensure the next generation. Leafy greens and root vegetables lack the fruiting structures to do this (tobacco and lettuce plants in particular are known to store perchlorate in their leaves). Which means poor Mark Watney in *The Martian*, despite his clever soil conversion work, probably would have died of perchlorate poisoning living off potatoes grown in the soil; at the very least he would have done serious damage to his thyroid.*

Various methods have been proposed to deal with the pesky perchlorate problem. It is a salt, so theoretically you can just wash it out of the soil. But the practicality of doing that where there is little water and the soil doesn't like accepting water to begin with is limited. A 2013 study suggested mining the perchlorate—taking the oxygen atoms to breathe and having innocuous chloride remaining. In 2016, students at Leiden University equipped *E. coli* with a set of genes from *Dechloromonas aromatica*, a type of bacteria that can break down perchlorate into chloride and oxygen.

In Wieger's lab they don't add perchlorate to their Martian soil because of its toxicity. Instead they add only chloride, at a rate you would expect to find with perchlorate, about two per cent. The aim is to see how that level of salt affects plant growth. Different plants have varying tolerance to salty soils

---

* Perchlorate had not been discovered in Martian regolith in 2011, when Andy Weir wrote *The Martian*.

but in general the saltier the soil, the harder it is for the plant to take up water.

At the time of our interview, they had only tried salt-tolerant plants including glasswort (*Salicornia*) and sea lavender (*Limonium perezii*). The glasswort grew very well—better with the extra chloride compared to the control plots—and had the wonderful benefit of lowering chloride levels in the soil while growing an edible plant. I've not eaten glasswort, but Wieger assures me you can taste the salinity in the plant, which is not a bad thing for those who like salty snacks.

5. Rest at 20°C for six months

Converting Martian regolith into soil will not be a fast process, but it is possible. Add compost, water, worms, microbes and other microbes to deal with the perchlorate, then grow some salt-loving plants to mop up the salts, and in a few months we might just have a Martian 'soil'.

What grows well?

Beans, tomatoes, kale, onions, lettuce, garden cress, dandelions, even hops—good news for Martian brewers—are common success stories. Everything Wieger tried eventually grew, though none of the plants grow quite as well as they do normally. He puts this down to the poor quality of the soil, even after being amended.

Potatoes—the crop most people ask about first, thanks to *The Martian*—have had varying success. The trick is using big enough pots and choosing the right potato variety in the first place. The ones grown by Mark Watney look like russets, but various attempts to grow them in Martian analogue soil have had mixed results. In some experiments the soil was too dense and the potatoes got squished. Wieger started with the Bilstar variety but was quickly put on the right track by potato farmers.

'I presented the results at a potato conference and all of them immediately knew that we had used the wrong variety to start.'

Those friendly potato farmers gave him Frieslander potatoes to use in his experiments, a better variety in their opinion, which did grow well.

Growing peanuts and lupins was harder, but essential for Wieger to try.

'I would like to have French fries when I go to space, and that means growing something with oil,' Wieger says.

Getting the peanuts to even germinate was a struggle. A few peanuts eventually grew but not enough to press for oil. In my personal experience peanuts are gluttons; they need high temperatures (they are a tropical plant), lots of water and lots of food. Low-nutrient soil in the Netherlands (or on Mars) is not an ideal environment.

Spinach was difficult but that was an experimental design issue. Spinach needs shorter days to produce lots of leaves (spinach prefers day length less than fourteen hours), but the greenhouse had automated lights, on for sixteen hours, off for eight. The long days meant the spinach flowered. Once it does that, the leaves become quite bitter and the plant won't produce many more leaves. Flowering spinach was a mistake but a positive one: a reminder that Martians will need the full plant lifecycle—flowers and seeds—to produce further crops. Most flowers need pollinating. There were no pollinators in the experiments until the flowers appeared and Wieger put his plant breeder training into action, hand-pollinating the plants with a paintbrush. This method works on an experimental scale but is too laborious to be feasible at commercial scale.

Wieger's pollinator of choice for Mars is bumblebees. Not only because bumblebee is a word he enjoys saying (and that sounds delightful in his Dutch accent, like 'bomblebee') but because they live happily in greenhouses and are good travellers. You can put a colony of bumblebees into hibernation for about six months, about as long as it will take to travel to Mars, then wake them up when they get there. Honeybees, however, did not survive in the greenhouse and are active throughout winter.

'Imagine having to bring a hive of bees with you into space with no gravity,' he says with a grimace.

Bumblebees pollinate the bigger flowers. Martians will also need pollinators that can work with smaller flowers. Australian native bees (my favourite are the blue banded bees with their black and cobalt blue stripy bottoms) can work with smaller flowers, as can flies—though if there is a chance to 'start again', would you really want to take flies?

Whatever the pollinators are they will need to be happy in a closed system with odd lighting schedules. Many years ago, I visited a research facility in the north of England where they were growing cannabis in a completely closed system using various wavelengths of red and blue light. The idea was to remove the psychoactive effects while keeping the 'munchies' and thus create medicine for cancer patients who felt too ill to eat. With no natural light you quickly get a headache, so we donned yellow glasses before heading inside. A few bumblebees buzzed sluggishly around the towers of pots, while others kept bumping into the ceilings and walls. Working out a way to keep the pollinators happy will be just as important as growing the plants.

As for the taste of Martian vegetables, multiple studies have found little taste variation between crops grown in Earth soil and amended Martian regolith.

To Wieger, the Martian crops had more flavour. 'The tomatoes, they are very nice and sweet. That is not because of the soil (I think), but because we do not grow the crops as fast as normally is done. And slower growth means normally more taste. Especially the spicy crops are more intense, for example, garden cress and rucola [rocket].'

Beyond plants and 2D farming

Growing plants is an excellent start to establish life on Mars, but agriculture overall will need a serious rethink. Everywhere inhabited by humans and plants on Mars will need to be contained, powered and pressurised; farming indoors, only on the ground, is a hectic waste of space. On Mars, food production will need to step away from the soil—perhaps towards vertical farming, already a common element of urban agriculture. Vertical farms are often hydroponic or aeroponic, circumventing the need for soil (though with hydroponics you do need some sort of medium to anchor the plants in) as well as a complete system of trays, pumps, reservoirs and so on.

These systems are space efficient, but more fragile than soil-based production. Soil has its own store of nutrients and water, so crops can survive for a few days if they are neglected. It can also buffer an accidental increase or decrease in fertiliser or change in quality. In hydroponics, nutrients and water tend to cycle through the entire system—if a pump breaks, water and nutrient supply stops to all crops. It can also mean havoc if a virus or disease gets in: disease and contamination spread very quickly through the recycling system.

I think hydroponics will be a necessary part of food production on Mars, but not the only form of food production. And in any case, we'd have a chance to reconsider the very notion of food: the word has a much broader scope when considered in the academic or literal sense, as something that people and animals eat, and plants absorb, to maintain life.

Everything living is made from the same twenty or so essential elements. Change how they're put together and these elements can become every protein, cell and amino acid that make up the billions of different species on Earth and the countless many millions now extinct. Food is something that provides these elements, or nutrients, you need. A penguin is food. Hay is food. Any fuel source that will sustain you is food—once again, imagine you are lost in the wilderness: the hungrier you get, the more widely you will consider what constitutes 'food'.

On Mars, nutrient-dense protein sources could come from algae, insects and lab-grown meat. Researchers at Penn State University proposed an ingenious system where you use your excrement to feed a 'microbial goo', which you then eat. It reportedly tastes a bit like vegemite. And while a vegemite and toasted cheese sandwich is a brilliant hangover comfort food, I think I'd reach for the plate of fried crickets first.

Insects—lightweight, low input and nutritionally comparable to traditional meat like beef and chicken—are an attractive food source, and hardly outlandish, given that they are currently consumed by about two billion people worldwide. Crickets and mealworms, for example, are incredibly rich sources of protein, amino acids, and some vitamins including B12. The comparison data varies, but roughly speaking a hundred grams of beef and a hundred grams of crickets each contain about twenty grams of protein. What makes insects the clear winner is the comparatively small expenditure of resources needed to obtain said protein. They can be fed food waste and are very space efficient. Circle Harvest in Western Sydney produces ten tonnes of insects each week on about half an acre. They're certainly not using the thousands of litres of water necessary to produce beef protein. And it's quick: you can go from insect eggs to cricket-meal 'corn' chips in about eight weeks.

There's also the possibility of cell-meat: meat grown in a laboratory. When you read about it in the press it is often accompanied by a photo of minced meat in a Petri dish, which is only going to look appetising if you haven't eaten anything for about a week. It is also not how you make meat in a lab.

You take a biopsy (a sample the size of an almond) from any animal or fish, break it down to isolate the different cell types (muscle, fat and so on), then put them in big steel tank called a bioreactor that contains growing media to feed the cells with the right nutrients, vitamins and minerals. Once the cells have grown, they are put together to make the 'meat'. At this stage, minced meat is easier—creating a steak with the right texture and mouth-feel is very complicated.*

Cellular agriculture is not limited to meat. Any animal product—dairy, eggs, even leather—is possible. Because you only need a small biopsy, you can eat things it is considered poor form to kill (Galapagos turtle, which Charles Darwin found to be particularly tasty, may come back onto the menu). But the most interesting part to me is that you can alter the nutritional content. There are opportunities to create hybrid meat-plant products, perhaps making a higher-fibre meat, or lowering the amount of saturated fat and upping the omega-3 in a steak. One research lab in the US is working on getting muscle cells to produce beta-carotene, which is normally only found in plants.

Obviously techniques like insect farming and cellular agriculture that don't use the soil directly are a little off-topic for a book about soil, but the point is that feeding Martians will require a combination of food-production processes. Maybe

---

* At the time of writing, Aleph Farms in Israel had recently managed to 3D-print a steak complete with fats, marbling, tendons and gristle. To my untrained eye, it looks just like a steak.

some crops grown in Martian regolith turned into soil, along with hydroponics systems, aeroponics, insects for protein, algae and cellular agriculture.

There are still many unknowns. Will plant and soil-borne diseases exist? Might new ones evolve? What new plant varieties might evolve? Fresh watery vegetables like lettuce might only be a luxury—they are big space takers with little comparative nutrition value and might only be a treat afforded to the very, very rich. (In the 1700s pineapples cost the equivalent of eleven thousand Australian dollars each. Maybe Martian pineapples will be similar.) Presumably even our feelings about food will be transformed. Psychologists say we need fresh food for mental health reasons: the familiarity and comfort it provides. But can you crave something you've never eaten? If children born on Mars only know insects and lab-produced food, will they really want to eat an apple?

And of course, none of these processes is applicable only to a future that involves extraterrestrial expansion. They all have implications for the future of life and food production on this planet, too—as do current experiments in building techniques.

'Earth' building

Even before food production, the first challenge on Mars will be building, because sending building materials to Mars is not feasible. Future Martians will have to rely on some of the first building techniques employed by humans: building with soil.

Historically and globally, soil has often been the easiest

and most convenient thing to build with. The original Great Wall of China was built (from about 221BC on) from stone (where it ran through the mountains) and rammed earth (when it crossed the plains). The oldest freestanding mudbrick structure in the world, at nearly five thousand years old, is the ceremonial enclosure of Khasekhemwy in Egypt. The Great Mosque of Djenné in Mali, built in 1907, is the largest. And of course animals, particularly termites, were building with soil long before humans appeared on Earth. Using only soil, water and their saliva, they can build mighty double-walled structures that can last for more than a hundred years and win the occasional battle with a wayward car.

There are many ways to build with soil. Adobe is made by sun-drying clay bricks. For wattle and daub structures wet clay, sand, animal dung and straw are mixed together and plastered over a wooden lattice. Cob construction uses layers of soil and straw and is useful for curved walls and arches. Rammed earth is my favourite: it makes some truly beautiful structures of sleek polished surfaces with layers and layers of ochre colours. It is made by ramming soil between two flat panels, usually to make a wall. There are many more methods with hundreds of variations on material ratios, soil types, compaction methods and drying methods.

Today, building with soil is often an environmental or aesthetic choice. Architect Giuseppe Calabrese explains why building with soil makes sense on Earth.

'It is pest-proof, you don't get termites, you don't get fire and it is excellent for [protection against] radiation,' he says. It is also pretty good at noise-proofing, helps purify the air and discourages condensation, even in bathrooms that usually get mould. Then the architect in him takes over: 'It creates a dialogue with the outside. There is a general sense of wellbeing when you are inside an earth building. It feels healthy. It's the smell of soil; it's different.'

Giuseppe began thinking about Martian regolith as a construction medium during some pandemic-induced downtime. Having organised all the files and archives in the office, Giuseppe convinced his company to enter a competition to design an urban farm on Mars. Their winning entry, 'Sprout', is a fractal design: from above it looks a bit like a snowflake with tunnels that radiate out from a central circle, with more tunnels branching off the main arms. The sides of the tunnels are specified as fifty-centimetre-thick mudbricks made from smectite clay harvested from the Jezero crater on Mars. In contrast to Earth buildings, Sprout does not try to create a dialogue with the outside; these thick mudbricks aim to protect against the deadly radiation. The domed tops of the tunnels are 3D-printed with a mixture of basalt and silicon extracted from rock. The silicon can also be used to make glass. It won't be beautiful clear glass, but it will let sunlight in. Mars doesn't get much sunlight; what it does get is too valuable a resource to waste.

Giuseppe's mudbricks are like giant Lego, big rectangles with points on top so the bricks can stack together easily. He proposes to make the mudbricks by simply smooshing the smectite clay under high pressure to make compressed bricks.

It's a technique that would not work on Earth. To start, Earth bricks need more ingredients to be stable and strong enough for use.* Second, smectite is a shrink–swell clay. On Earth, if you used more than a small amount of smectite the bricks would swell when damp and crack when they dried out. But clays don't swell without water, so if there's no water on Mars, this might not be a problem (providing the bricks aren't exposed to more humid conditions indoors). When Californian researchers made compressed bricks out of simulated Martian regolith, they ended up with bricks as strong as concrete. Granted, their bricks were much smaller than Giuseppe's giant Lego, but it's a start.

It helps that Martian regolith is naturally prone to compaction and tends to set hard, a property NASA's InSight lander is only too familiar with. It was supposed to dig a five-metre hole to measure internal temperatures to give scientists clues about how the planet formed and has changed over the last 4.6 billion years. Unfortunately, the 'mole', as it's nicknamed, needs friction from soil flowing into the hole so it can move around. Where the mole tried to dig is a layer of duricrust, compacted soil that won't fall back in as the mole works. In early 2021, almost two years after landing, it was still stuck. Buzz Aldrin encountered a similar issue on the Moon when trying to stick the American flag in the soil, experiencing momentary panic that the flag would just fall over.

---

* Earth bricks use a mixture of different particle sizes (sand, silt and clay) with straw or manure as a binding ingredient. Modern bricks are often mixed with concrete or lime to give them extra strength and to meet building codes.

What helps the regolith set hard is iron oxide, the compound that gives Martian regolith its red colour. When crushed, iron oxide forms very strong bonds—it's the same stuff that helped the Cu Chi tunnels in Vietnam withstand explosions from US grenades (War chapter). Alas, as on Earth, Martian regolith is not uniform and iron oxides are often found only in a thin layer on the surface. Dig down and there are at least twenty-nine other colours of regolith ranging from blue-grey to red, along with light yellows and browns, indicating different minerals that might not make such sturdy bricks.

Where simply compressing regolith is not sufficient, Martian bricks will need some sort of binding agent to add strength. Many institutions are investigating various agents such as sulfur, chitin (extracted from insect and crustacean exoskeletons), and even blood and urine.

So, as on Earth, no one building process is 'the answer'. The variability of Martian regolith means the exact formula or mix of regolith and binding agent could vary from place to place. Martian colonies might resemble older places on Earth, with the materials based on what was available at the time and what construction method suited those materials.*

Because of the frequent dust storms, there might be areas

---

* 'In Italy it is evident,' says Giuseppe Calabrese. 'If you were in an area that has a whole lot of clay all the buildings are made from mudbricks. If you are in an area with a lot of limestone the whole town is made out of limestone. It makes the sense. But now you can go to Tokyo or Sydney or the US and every building looks the same—it's all made out of the same material.' His look of disdain as he says this is...architectural.

on Mars where iron-oxide–rich soil has piled metres and metres deep—this is where locally made Mars bricks might be fashioned. Other areas might need a different style of brick. The exact formulation might also play a role in building design. Some areas might have blockier structures like Giuseppe's farm Sprout. Other areas might have buildings more akin to what we imagine as 'space age'—domes and cylinders and odd curvy structures—if we can get the tech right.

Giuseppe's practical approach, as well as the hundreds of pages of scientific research that backed up his design, played a role in him winning the competition. A few scientists on the judging committee commented that finally, they'd been presented with something they could construct.

Giuseppe refers to some of his competitors' designs as 'Zaha Hadid* structures...tricky shapes, but just not realistic. They are what we call stage three architecture'—nice in theory and aesthetics but lacking practicality.

'What I focused on was something that could be built,' he says. 'Even though it is not that aesthetically pleasing.' This is probably difficult for an architect to admit, but for what it's worth I disagree. Sprout looks fine to me.†

On Earth, curves are tricky and expensive to build, which is partly why curved structures are not the norm. But they are not impossible and might be more common on Mars

---

* Zaha Hadid was an architect once described as the 'Queen of the Curve'. Take a look at the Heydar Aliyev Centre in Baku, Azerbaijan, for one of her particularly curvy designs.
† Caveat: I don't have a discerning eye when it comes to architecture.

than they are on Earth, especially if buildings are made by 3D-printing.

3D-printing with Martian regolith

In early May 2019, AI SpaceFactory and Penn State University faced off in the grand final of NASA's 3D-Printed Habitat Challenge. Over three nail-biting ten-hour days, with prize money of half a million US dollars at stake, the teams watched their robots extrude layer after layer of brown or grey goop as their designs for housing on Mars grew slowly upwards. Since 2015, the teams had submitted house designs, printed various structural components of the buildings and refined the 3D-printing technology. The final stage of the competition was to print a one-third scale model of their designs. For AI SpaceFactory it was the first time they had tried to build a complete version of their design 'Marsha'—until then, they had only printed pieces—and as the clock counted down the final few minutes all was going well. Then AI SpaceFactory's robot placed the finishing touch on their large, dark brown cylinder: a circular glass panel that sat on the very top. The smiles of relief were short-lived. A few seconds later there was an almighty crash as the glass fell through and smashed on the floor.

With 3D-printing, curves are easier to make than sharp-edged blocks, which is perhaps why so many designs in this competition, including the two finalists, were curvy. While AI SpaceFactory went for the cylinder, the Penn State design looked a bit like gnome houses.

3D-printing is an attractive option for Mars because it significantly widens construction options. It can produce tricky shapes like lattices that are lightweight and strong, using precisely the quantity of materials it needs. With 3D-printing there are no offcuts.

For the process to even be considered an option, Martian regolith needs to be a key ingredient, if not the main one. AI SpaceFactory used a mixture of basalt (notionally extracted from Martian rock) and a bioplastic made from plants—a mix that can be reheated, melted and reprinted into something else. Penn State developed MarsCrete™, a mixture of basalt rock, kaolinite, sodium and silicon, all of which can be harvested on Mars.

As with brickmaking, getting the blend right might be the trickiest part. Because, as we've seen, Martian regolith is not uniform, so the method needs to work with varying particle sizes and minerals, and with the binding agents necessary to add strength. The consequences of mistakes or poor formulation would be dire. Ideally, the equipment would be able to analyse the regolith as it was extracted, and automatically adapt the composition and quantity of the binding agent as printing progressed.

Both of the competition finalists were subjected to structural tests—which in this case meant trying to squash them with a nine-tonne excavator. The Penn State gnome house crumbled but Marsha held strong: the excavator began to lift up at the front as it tried to crush her with its bucket. The combination of basalt fibres and polymer also reportedly outperformed concrete for strength, durability and crush testing, and was five times more durable in freeze-thaw conditions—pretty important considering the night temperatures on Mars.

Despite their smashed glass panel, AI SpaceFactory won the competition, taking home the five hundred grand and a few extra grey hairs.

Death

The cost of grass

Dying is a very inconsiderate thing to do: inevitable, but inconsiderate. On top of the grief you've caused family and friends, now there's a great big bill—and a corpse to deal with.

Bury or burn are the two most popular choices. They have persisted throughout history, with religious factors largely dictating which one. Traditionally, Muslims, Jews and Christians bury (and may even prohibit cremation), while Hindus and Buddhists burn. Which is why, until recently, burial was the more common choice in the West. But things are changing. In Australia cremations now account for about sixty-five per cent of end-of-life choices. In the USA, the burial rate was surpassed by the cremation rate in 2015, and is predicted to be down to thirty per cent by 2025.

Is this swing in end-of-life proceedings because the world is becoming more secular, or do most people now simply prefer the idea of cremation?

'It's cost,' says Kevin Hartley, who has been a funeral director for over twenty years. We've sat down to morning tea at his house in Melbourne. Kevin sits like a yogi, cross-legged on his chair, while explaining the finer points of managing funerals.

'Burials are at least double, usually three times the price of a cremation,' says Kevin. 'Even if someone wanted to be buried, when I mention the price to the family, there's some-times an awkward shuffling of feet, some clearing of throats, then the family decides on cremation.'

The average cost of a burial in Australia is just under twenty thousand dollars, more than the average annual pension. It seems like an outrageous amount of money until you look more closely at what's actually involved with a burial and funeral. It's like a four-dollar cup of tea in a café: you're not just paying for the tea. Kevin rattles off a list of expenses: the funeral director, staff to organise the funeral, logistics, transporting the deceased, temporary 'accommodation' for the deceased until it's time for the funeral, the back-up trans-port in case the hearse breaks down (things are not allowed to go 'wrong' at funerals).

But the burial plot is usually the most expensive part. For some people the plot might be the most expensive real estate you'll ever own (though in some cases 'own' could mean rent, since what you might be paying for is a 'right of interment', which is the right to be buried on that plot of land for a certain amount of time). At the upper end, a plot in the eastern suburbs of Sydney will cost nearly fourteen thousand dollars. Graves are about 3.5 square metres in size, making that about sixteen million dollars per acre. Cheaper plots, at about four to five grand each, are equivalent to about five million an acre.

'Then there's the coffin; sixteen to eighteen hundred dollars to dig a grave—'

At this point I hold up my hand. For most of the prices Kevin has given I have no real frame of reference, but I know what it costs to get a backhoe to dig a six-foot hole. It does not cost sixteen to eighteen hundred dollars.

'Because it's a profit point,' Kevin explains. 'You try to have multiple profit points so none of them look too extraordinary.' Like the two-hundred-dollar chipboard coffin with a stick-on veneer that sells for fourteen hundred.

It's an odd sensation, finding 'profit points' more uncomfortable than the topic of death. Yet Kevin assures me that funeral homes (mostly) aren't trying to take advantage of the grieving. Cemeteries need to find these profit points because of the cost of land and green grass. 'There is an implicit understanding that when Grandpa is buried at that plot in the lawn cemetery, that patch of grass will be maintained forever,' says Kevin.

I spent the first five years of my career digging holes on sports fields and Kevin started out as a greenkeeper. We munch on cake in silence, contemplating our mutual familiarity with a turf-maintenance schedule. Fertilising, weed spraying, irrigation when it's dry, mowing and whipper-snippering the long grass around the gravestones. Green grass, especially on the second-driest continent of Earth, is an expensive luxury. If society expects a tidy green lawn in perpetuity, it must be paid for somehow.

'During heavy rain,' Kevin says, referring to a time when he worked at a cemetery that had flat headstones, 'three guys did nothing for a week except whipper-snipper around the headstones.'

Society does not tolerate a messy graveyard. Those staff need to be paid, and the costs add up. The cemetery proposed lowering the headstones by ten centimetres so they could go over them with a mower to save time, and touch up with a whipper snipper. But public outrage prevented that course of action. Sometimes we seem to have more concern for patches of earth where we bury the dead than for the land that keeps the rest of us alive.

In this busy modern world, lawn cemeteries are an inefficient and expensive way of doing death. Large countries with small populations, like Australia, will always be able to find space for burials, though it might mean a longer commute to the cemetery (Sydney's metropolitan gravesites are predicted to be filled up by 2051). But in smaller, land-starved countries, a lawn cemetery is unthinkable: an epic waste of resources. In Singapore, the Choa Chu Kang Cemetery Complex is the only cemetery still open for burials. In Hong Kong the cremation rate is around ninety per cent, but with thousands of containers of ashes stored in funeral homes, space is starting to get tight even for these.

Reusing graves is one way to combat the space issues. The practice might initially seem unwholesome, but is quite common in land-limited countries. In Greece, for example, on the third anniversary of a burial the person is exhumed and washed with wine, and the remains are placed in an ossuary (a special container or building). This allows frequent grave reuse in a country that simply would not have the space to bury everyone who has died in Greece in the last nine thousand years. The tradition respects the remains of the dead while recognising the importance and limited availability of land.

In Australia, it is a good idea to check* whether your plot in the cemetery is yours in perpetuity or for a specified time-frame, after which the plot needs to be re-leased or it will be used by someone else. Once your lease is up, the family or descendants are contacted to re-buy it. If they don't, there is a process called the 'lift and deepen', which has been used for many years in European cemeteries. Cemetery owners remove the headstone, dig the grave again, put the bones in an ossuary at the bottom (or move the bones elsewhere), and prepare the grave for the next family.

How soon one might reuse a grave has come up for discussion in reviews of the Cemeteries and Crematoria

---

* The policy varies between states and, because it is a dynamic point of discussion, may change between the time of writing and publication of this book.

Amendment Regulation for NSW. One suggestion was as little as twenty-five years.

'You should not reuse a gravesite within fifty years,' says Boyd Dent, hydrogeologist and expert in cemetery soil and water management.

The recording of my interview with Boyd is hard to hear. Partly because we're in a crowded café with a background din of chattering, a coffee machine and clinking cutlery; partly because we're trying not to speak too loudly about the logistics of cemetery management when surrounded by predominantly grey-haired patrons enjoying a nice cup of tea and a slice of cake. Boyd has studied the soil at enough cemeteries and seen more than enough exhumations to have a firm opinion on the matter.

'Ninety-nine years is better. You are asking cemetery workers to dig up basically human goo after twenty-five years. That's a very difficult task.'

Or as Kevin put it, rapping his knuckles on the wooden table and making our teacups clink: 'It's a very distinctive noise when a backhoe bucket hits a coffin.'

At the time of writing, legislation is a little vague on what state a body should be in before the grave can be reused.

'It would be better if they said skeletonised, which would avoid ambiguity,' Boyd suggests.

'But how would you know it had reached that state without digging it up?' I ask.

'I'd like to know that,' Boyd says. 'It could be a lot more than twenty-five years.'

How long it takes a corpse to become a skeleton depends on the climate, the soil and how the person was buried. Someone in a shroud buried in sandy acidic soil with good drainage will decompose much faster than someone in a coffin buried in clay. Coffins seal off the body from the soil, especially coffins lined with plastic. In an anaerobic box, natural decomposition (what would be happening if the person was in the soil) is severely slowed down because the insects and microbes simply cannot do their work. In time, the coffin degrades somewhat and collapses, but this can take decades. Embalming adds even more years to the process.

Boyd has witnessed the exhumation of a man who had been buried in a coffin in a clayey soil for about fifty-six years.

'He was skeletonised'—Boyd pauses to take a sip of his coffee—'except for his brain.'

Not knowing how long it takes to become a skeleton is a big impediment to grave reuse, especially for cemeteries situated on clay soil in higher rainfall areas. It is enough of a problem that Rookwood Cemetery in Sydney is running a special trial known as the Rookwood Experiments, which sounds like a horror movie, and has elements of such. Ninety-three pigs* have been buried using a range of burial techniques (above and below ground) and coffin types, with a goal of better understanding cemetery soil management, including how clothing, coffination and different types of coffins affect decomposition.

---

* Even though Rookwood is a cemetery it would be illegal to bury human bodies there for research. The only place you can do that is the Australian Facility for Taphonomic Experimental Research (AFTER).

The longest that specimens will be interred for is five years, finishing in 2023. Unfortunately, this book will go to print too soon to use any data from that work. But Boyd informs me that the experiments have so far yielded a patented above-ground 'burial' process designed to provide burial space in areas where the land isn't suitable for in-ground interment—perhaps because the soil is too shallow or the watertable too high.

In this system, concrete burial vaults are stacked above ground, containing soil with the right properties for optimal decomposition. A core of soil runs through the vaults, too, in deference to religious traditions in which the deceased must always be in contact with the earth. The decomposition products (body fluids and gases) are adsorbed, absorbed, transformed and permeated through the vault's soil drainage bed and soil core. The plan is that after skeletonisation the vault is opened, the bones are placed at the back of the vault to make space for the next person, and the vault is resealed. Results so far show decent decomposition speed—at least faster than burial in the local clay at Rookwood.

Of course, decomposition isn't the only issue. The social aspects of plot reuse present obvious challenges. And financially? A shorter, renewable lease on a plot of land could mean families end up paying even more for a gravesite.

For now, populations are increasing and more people means, ultimately, more dead people. Highly competitive and expensive burial plots mean fewer can afford that option and cremation is picking up the slack. Which presents its own soil issues, particularly when it comes to scattering ashes.

The ash problem

'I want to be cremated and spread across my rose garden,' Mum says whenever the mood strikes to discuss her eventual demise. Death is not a taboo topic in our family. 'I don't want there to be any chance that I'm still alive, so burn me.'

I suggest we bury her with a little bell on her big toe like they used to do, just in case. But she wants cremation.

Cremation is a very efficient way of shrinking the burden that is a corpse, turning an eighty-kilo human into about 2.5 kilos in just a few hours. Some cultures have done it for centuries, burning their dead on a pyre. Today we have a system: corpse to container in as long as it takes to bake a cheesecake. Ninety minutes at nine hundred degrees Celsius just about does it. Wait for the ashes to cool, pop them in an urn, and you've got Grandma strapped in the back seat of the car.

But then the question arises about what to do with the ashes. My grandmother's ashes sat in our linen press for many years, while my parents joked about taking her to school when family-history assignments were due.

Many people, just like my mother with her rose garden, want to be scattered somewhere beautiful. It is a lovely sentiment. But the reality—in fact the chemistry—is that scattering too many people in beautiful places wreaks havoc on the environment.

Your ashes—that two and a half kilograms that remains after all the fats and organic matter are burnt off*—

---

* The method of cremation is not too different from how soil scientists can test the amount of organic matter in a soil sample. Weigh out a sample of

will be unique to you, the exact percentage of elements depending on your lifestyle and history. If you lived in an area that gets acid rain, you are more likely to have more copper, lead and cadmium in your ashes, as the lower pH of the water increases their absorption. Vegetarians are more likely to have higher levels of strontium, of which plants are the best sources.

But basically, human ashes are mostly bone, which means they are predominantly carbonates, calcium (about twenty-five per cent) and phosphate (about forty-eight per cent).

The levels of calcium and phosphorus, as well as salt, can make ashes a horticultural issue. A little scatter here and there isn't a problem, but if thousands of people want their ashes scattered in the same place, soil chemistry starts to change. Most plants prefer the slightly acidic pH that makes nutrients more available for uptake. Ashes are quite alkaline, so adding ashes is like liming the soil. Soil pH starts to rise, and some essential plant nutrients such as iron, manganese, copper and zinc become harder to access. The calcium and phosphorus aren't initially a problem, as they take a while to weather and become available for plant use. But again, if too many people choose the same location, it's like dumping phosphorus fertiliser on the ground. Not all plants like lots of phosphorus, and changes in the soil chemistry can encourage less-desirable plant species.

---

soil, put it in a special oven at about six hundred degrees Celsius overnight, then weigh it again the next morning. The heat burns off all the organic matter, making the difference in weight the amount of organic matter in the sample.

The other issue is that people tend to put ashes in a mound in one place instead of scattering them around. Concentrating ashes in one spot is better ceremonially, but not horticulturally—nor psychologically. Surrounding plants can get scorched and the mound can become a 'sludgy' mess.* Piles of ashes have intermittently become a problem in many beautiful places in the UK—the Lake District, Dartmoor and Scott's View near the Scottish border. The gardens at Jane Austen's house in Hampshire started to wither when too many mourners scattered loved ones on the grounds. The practice is now banned—for the sake of the garden, the gardeners, and visitors who don't like seeing mounds of human ashes while walking the grounds.

Scattering ashes can be a lovely end-of-life ceremony if done properly. Expert advice includes scattering below waist height, standing upwind, and not upending the container of ashes unless you have a rake. (There will be more than you think, and having to level out the mound with your boots probably isn't the look you were going for.)

A few enterprising companies have come up with unique ways to deal with ashes. You can become a firecracker or bullets, or a set of coasters. You can be turned into a glass vase or even an hourglass (you won't be an accurate timepiece, but your family will get a constant melancholic reminder of how

---

* As described by Scottish solicitor William Windram, urging the local crematorium to give advice about disposal of ashes to preserve the Rhymer's Stone in Melrose, Roxburghshire. Quoted in Ruth Warrander, 'Mourners turning Borders beauty spot into "sludgy" mess by dumping loved ones' ashes', *Scottish Sun*, 3 May 2019.

quickly life drains away). Becoming a pot plant is popular, after the ashes have been treated to make them more suitable as a fertiliser and less hostile to plant roots. Some manufacturers even make urns that contain a soil moisture sensor to tell you when Grandma needs a little water, and sensors that let you know if nutrient stocks are running low and it's time to fertilise.

Being ecofriendly when you're dead

It is because of the space issues, the cost and the ash problem that Kevin Hartley started thinking laterally about his craft. After decades of doing funerals the 'traditional way', with coffins, embalming and lawn cemeteries, he wondered if there might be a different way to deal with death.

'Crematoriums are a crude industrial incineration process. Just another contribution to CO_2 gases,' says Kevin. The average US cremation creates around the same emissions as two tanks of petrol. Cremation also releases the mercury from anyone who still has mercury dental fillings.

And many people, regardless of faith or religious views, still want to be buried. There is something that resonates about being returned to the earth. 'I used to work with someone,' Kevin reminisces, 'who knew how to add just the right amount of flair to a [burial] service. He had a special silver case that looked a bit like a sterling silver lipstick case. When the minister got to the part about dust to dust he would step forward and, right on cue, scatter some clean sand onto the coffin with a sombre flourish.'

Those who have died return to a place; those still alive want to return to that location to remember them. Natural burials offer another way to do that. Most also aim to cost less than a modern burial in a lawn cemetery, and with less environmental impact.

Imagine a wildflower meadow, soft blooms rustling in the breeze. A gravel path meanders up and down the gentle slopes, passing beneath large shade trees dotted across the landscape. The native shrubs are in bloom, buzzing with bees. There are no headstones. The ground is dotted with small circular bronze plaques that you wouldn't notice unless you were looking for them. In many places the grass has grown over them, but it doesn't matter—each bronze circle is geolocated.

A funeral service is starting. It looks like a normal procession in every other way. Pallbearers in black suits bring the coffin to the gravesite. There is a ceremony, and the coffin is lowered into the ground.

Then the coffin is removed. Kevin has invented special equipment to retain the dignity required on such a solemn occasion. His transporter, which made it to the ABC TV show *The Inventors*, looks like a coffin but has a trapdoor at the bottom. The device makes moving a shroud-wrapped person with care and respect much easier. (Wrapping the body is a common feature of natural burials, a shroud being more eco-friendly than a standard coffin; unfortunately it doesn't have the rigidity needed to move a body gracefully.) At a different funeral, it might be a biodegradable casket or wicker coffin, instead of a heavy wood and metal box that takes years to degrade. Each element of the funeral and burial has been carefully thought out to offer the poignancy of a funeral, but with less ecological impact.

'No headstones,' Kevin repeats. Although we are used to seeing stone markers, they are troublesome in a natural burial setting. Grave markers in general are a problem, as people like to garden. The bigger the monument, the more people will plant or add trinkets around it, which somewhat spoils the idea of a natural landscape.

Designation as a cemetery confers a degree of protection on the land, along with opportunities to replant, create habitat and protect our dwindling wilderness. 'It's reverse mining,' Kevin says. Everything we have we get from nature and, according to Kevin, we've overdrawn the account. 'We take and accumulate when on this Earth. It makes sense we return our elements back to the Earth when we're done.'

Environmental impact concerns are a key driver of natural burials, because wooden coffins are not environmentally friendly. They are often prepared with varnishes or sealants that can contain arsenic, the glues usually contain formaldehyde, and many coffins come with plastic liners. Burial vaults are made from a variety of products that can leach products to the soil, including plastic, concrete and asphalt. Embalming is not mandatory in Australia, except in certain special circumstances, such as when the body needs to be transported, but embalming fluid is a noxious brew of formaldehyde and other solvents, preservatives and disinfectant agents. Another common yet less discussed source of 'contamination' is the pesticides and herbicides used to keep grass green and weed-free. If society can't handle broadleaf weeds mixed in with the couch and kikuyu, someone needs to remove them, and in this country that is not going to happen by hand.

The aim of natural burial, then, is a 'win-win-win', as Kevin puts it. For the family, the cost will be the same as a cremation, with the money contributing to preservation of the wilderness area and the cost of upkeep and expansion. For the land, the system rehabilitates degraded land where possible, reduces pollution, and protects at least some areas from development. For the atmosphere, Kevin is aiming for carbon-neutral funerals.

Kevin's is far from the only natural burial idea. There are nearly three hundred natural burial sites in the UK and a growing number in the USA. Many align with Kevin's views on death done better, cleaner and with an ethos of giving back to the earth. In Kevin's version, degraded land is returned to native vegetation, perhaps grassland with wildflowers and native shrubs.

Woodland burial sites are a nice idea, but although everyone wants to become a tree (or so it seems—it's very popular), it is simply not possible for every burial. Woodland burial sites usually don't offer large trees as an end-of-life marker, suggesting instead a selection of shrubs or wildflowers. It's not like the forestry industry, where you can plant large trees in close rows because they'll be cut down in twenty or thirty years. Row upon row of huge, beautiful trees that will last for centuries? Not possible. The roots get tangled, the trees compete for light, water and nutrients, and overcrowding means some will win while others will wither.

Have you ever thought about the traditional phrase 'ashes to ashes, dust to dust'?* It implies that we come from the earth and shall return to it—just at the point where we're doing everything in our power to protect our deceased person from nature and the elements. The beauty of natural burials is that they fulfil the traditional vision. They connect death to life, viewing death as a process rather than an ending, and they change how we view decay. They celebrate rather than abhor it, recognising that all living things are part of nature and, in time, will return to it.

I quite like the idea of natural burials. I'm more than happy for my organs to be reused if that's an option (though I would prefer to live long enough to thoroughly wear them out), then pop me in the ground in the paddock or meadow or woodland. My elements, rented out for decades to my consciousness, can return to the soil, pass into the land of the abiotic, then back into the land of the living once more. Maybe as grass, a flower, a tree…maybe in time all three. Humans have a very intimate relationship with soil, we often just don't realise it.

With society accepting and, in some places, demanding more ecofriendly interment options, other companies and

---

* Specifically the Anglican tradition, although it seems to have spread. 'We therefore commit this body to the ground, earth to earth, ashes to ashes, dust to dust'—from the Burial Service in the Book of Common Prayer.

institutions are considering alternative options for end-of-life. Recompose, based in Seattle, is the first legal human composting facility. Bodies are placed in a shroud, covered in straw and woodchips (a source of carbon to help the composting process) and aerated (and presumably turned), allowing microbes to get to work. I imagine the process is similar to garden composting, but without the pitchforks.

The process claims to turn a body—including bones and teeth—into soil* in thirty days. 'The material we give back to families is much like the topsoil you'd buy at your local nursery,' reads the website. A lot of soil: much more than the volume you go in with. A typical body generates about a cubic yard—about three-quarters of a cubic metre—enough to fill the back of a ute or pick-up truck. That's more soil than many people can deal with, so Recompose has a land trust you can donate it to.

Getting rid of soft material in thirty days sounds reasonable with properly controlled composting: sustained high temperatures and a good balance of water and oxygen. The microbes break down most pathogens (tuberculosis is one notable exception); and of course there is no need for caskets or embalming fluid.

But as forensic archaeologist Eline Schotsmans (see the Crime chapter) says: 'If it was possible to completely become soil [just by composting] in thirty days, forensic scientists

---

* I assume when they say soil they mean compost, a soil amendment—which is, obviously, what you get when you compost. Unless they are also adding loam or another soil to contribute the mineral portion of a regular soil recipe.

would be out of a job. It is very hard to get rid of a body,' and that's mainly because of bones and teeth.

I asked Recompose for clarification about the outcome, as well as a hint about the chemistry of the compost—is it alkaline? Salty? What ratio should I mix it in with regular soil?—but did not receive a response. Unsurprisingly, the process is hush-hush, for intellectual property and perhaps marketing (and forensic) reasons. But I know enough about composting to know that you can't get rid of bones in thirty days unless you're using acid or pulverising them—which might be what is meant when they say the bones are broken down by 'mechanical means'.

The other very intriguing end-of-life option is the infinity burial suit or 'mushroom suit'. A fancy type of shroud, the suit is impregnated with fungus spores and claims to cleanse the body and soil of toxins. Mycoremediation, a technique that uses fungi to decontaminate water or soil, can be used to break down or bind up a variety of environmental toxins including metals, pesticides, hydrocarbons and pharma-ceuticals. The key is having the right sort of fungi (there are millions of species) in the right environment to maximise their clean-up potential, and ensuring they can keep up with whatever the pollutants are.

Just some of the things I wanted to know about corpse-specific mycoremediation included what toxins the fungi are removing, how long it takes, what soil types has the suit been tested in, and if there are different suits for dif-ferent climates and soils. But alas, like human composting, the science is very hush-hush, and I could not elicit a response from the organisations I contacted.

A special kind of landfill

Boyd Dent's expertise in cemeteries started with an interest in landfills that morphed into a PhD in cemetery hydrogeology. As he sees it, cemeteries are a 'special kind of landfill': at their core, a bunch of organic matter buried in the ground and covered with soil. The organic matter degrades over time, releasing gases and leachates into the soil. One notable difference between cemeteries and landfills, however—apart from the type of organic matter that is buried—is that landfills are required to constantly monitor their emissions and off-site impacts. I did this for a few years, measuring gases coming off the surface of various landfills, collecting water samples from bores dotted across the sites and measuring water quality in the surrounding waterways. The aim was to make sure nothing nasty was leaching out of the landfill.

When a body decomposes it releases nutrients, a salty plume, and maybe some pathogens. But cemeteries in Australia do not need to check for off-site impacts, or even that the leachate from coffins or corpses is not moving into the groundwater or surrounding waterways. So how do we know what impact cemeteries have on the soil and surrounding environment?

We don't, really. Boyd's PhD thesis, the seminal work on the hydrogeology of cemeteries, was submitted in 2002 and there is still limited research on the matter. The information we have is largely from mass graves, disasters or forensics, the latter focusing on how the soil impacts the body, not the other way around.

Should cemeteries even have to monitor? Boyd thinks not. It would completely change the landscape, for want of a better term.

'I don't want cemeteries treated as a contaminated site. And the thinking behind implying that monitoring must be included is that they are somehow contaminated. Treating them as a contaminated site introduces a whole new paradigm regarding how you approach and manage cemeteries.'

The idea that corpses are contaminated and dangerous to the surrounding environment—particularly groundwater—stems from very real but different situations. Water contamination from graveyards has been an issue historically, usually after mass death events like cholera or typhoid epidemics, or mass disasters. The difference here is that a) masses of people were crammed into cemeteries and b) historically, cemeteries have been poorly located: close to watertables and drinking-water sources, on drainage lines, in swampy soils, or on the edge of cliffs. This information is not overly useful in a well-managed cemetery situation.

'My thesis has always been if a cemetery is properly sited and operated, the chances of off-site impacts are very low,' says Boyd. 'So don't bury in bedrock, you need a deep soil profile, you need a soil layer under the coffin and above the bedrock, you need to be above the level to which any groundwater table rises so it has to be seasonally looked at. And leave a buffer zone.' He recommends a minimum of about twenty metres between the graves and cemetery boundary; drinking-water wells should be at least two hundred metres away from a cemetery.

'If you follow good practice the risk is very low. But there are certainly places where you shouldn't bury.' Ensuring there is a solid buffer zone—and not letting graves creep into it, as sometimes happens—keeps the risk of off-site impacts low. If a cemetery is too close to waterways or the watertable, or the soil is too sandy, there might not be enough time for the soil to do its job. The UK has guidelines similar to those Boyd recommends, with buffer zones from waterways, minimum amount of soil between the base of the grave and the top of the watertable, not allowing graves in bedrock, and at least one metre of soil on top of the grave.

Properly siting and managing a cemetery helps the soil atten-uate the decomposition products—which broadly means the soil can, through physical, chemical or biological processes, reduce the amount or toxicity of contaminants. Attenuation is the idea behind most septic systems—that the soil will deal with the leachate before it has a chance to reach any water sources. Note, it is not wise to compare a cemetery more closely than that to a septic system, as the loading is very dif-ferent. As Lee Webster writes in an article for the Green Burial Council: *One body will decompose over a period of four to six weeks, releasing about twelve gallons [54 litres] of mois-ture. The average family of four flushes 250 gallons [1150 litres] of water every day with 'infectious' waste that possibly contains drugs such as chemotherapy or birth control.*

In a cemetery situation, ideal soils are those that facilitate the breakdown and uptake of decomposition products. Soils should be sandy clay to clayey sand and, if possible, slightly acidic. They need to offer enough oxygen to facilitate decomposition but are not so sandy that any rain or water leaches quickly through the soil and deeper into the ground or into the watertable. Oxygen is critical for respiring microbes to do their work, which is where natural burials can help—no coffin, or only a wicker coffin. Shallower burial also helps; so does wrapping the body in natural, biodegradable fibres—and no trinkets like phones or jewellery allowed in the grave. All this helps speed up decomposition.

Too-sandy soil is also an issue when it comes to actually digging the grave: the grave walls don't hold up well for the process of getting the body into the ground.

Low-oxygen, wet environments tend to promote preservation rather than decay. In 1950, the Danish police were called to a peat bog near Silkeborg where a family digging peat for fuel had unearthed what they assumed was a fresh corpse. But upon seeing the body (buried 2.5 metres deep in undisturbed land and 60 metres away from solid ground) the police called the museum. The 'Tollund Man' had lived in the fourth century BCE but was so well preserved that blue tattoos still stood out on his skin. He had failed to decay for the same reason peat bogs can exist: the climate is too cold and anaerobic for much microbial activity.

—

Where much more work is needed is in the realm of pathogens.

'My major concern about cemeteries is the escape of pathogenic bacteria and viruses,' says Boyd. 'That has been my position since my masters degree, and that still scares me a lot.'

In theory, once someone dies and is buried, any infectious pathogens they harbour will eventually die too. If these organisms don't automatically die because they no longer have a living host, they will die because they are now locked away in a coffin underground. In reality, we don't know. There is very little research on the matter. We do know that some pathogens like anthrax (a soil-borne organism to begin with) and tuberculosis are quite capable of sustaining themselves without a human host.

Coffins complicate things because they can act as a lovely growth media for certain pathogens to grow and replicate. 'We have definitely shown in the Rookwood studies that plastic liners delay decomposition and the coffins effectively become a bath that contains human soup, if you like. When the plastic liner finally degrades a little or detaches from the coffin, the leachate is then free to escape into the environment,' says Boyd.

The assumption is that the soil deals with the problem. Which it may very well do—but how much and how fast? And how differently in all the different soil types, especially at 1.8 metres deep (six feet under), where many graves are and where there is less microbial activity than at the surface? And what happens if we start reusing graves every twenty-five years?

'It's a big thing that should be investigated, and if you give me sixty million dollars I can set up a wonderful research project to answer it,' Boyd says.

I'd like to be part of that research project, even though it would be very hard to do. Work like this requires a great deal of sensitivity and a strong stomach, having to dig under graves and do exhumations. Aware of this, Boyd has suggested we at least dig up the smallpox victims at the quarantine station in Manly to check how smallpox has survived. It's not an idea that has been taken on.

One review paper from 2015 summarised research from around the world (including Australia, Brazil, South Africa, Portugal, the United Kingdom and Poland) on groundwater contamination by bacteria and viruses. Based on the studies available the authors concluded that the risk of groundwater pollution by bacteria and viruses is low in 'moderate climate conditions', with risks rising in tropical climates, but arguing that these risks can be managed by properly siting cemeteries to begin with.

So overall, I remain positive. If humanity has survived this long without knowing the answers, and in situations like Greece where reusing graves is common practice, we will probably be fine. Moving to more ecofriendly burial practices that facilitate rather than impede decomposition should help keep risks low.

But as a scientist, I would like to see some more data.

War

THE HISTORY OF war is littered with examples of terrain, particularly soft ground and mud, exerting a catastrophic influence on the course of battle. Waterloo, Agincourt, the Kokoda Track campaign—and World War I in particular: the Somme, Passchendaele and Verdun. Probably the horror that is most viscerally remembered about the conflict on the Western Front—apart from the sheer scale of the casualties— is the mud. In these battles, torrential rains and sustained bombardment fell on heavy clays, turning fields into quagmires. Horses got stuck, rifles clogged, medical help was hindered. Shell holes became treacherous swamps, trenches collapsed on the men they were sheltering—soldiers drowned in the mud, a fate thought to be worse than death by bullet. As front-line newspaper *Le Bochofage* said: *Hell is not fire,* *that would not be the ultimate suffering. Hell is mud.*

Soil is, naturally, part of almost every war. Wars are often fought over the soil itself—the land—whether to expand an empire or capture resources beneath. Sometimes the soil turns to mud; sometimes it is dust that is the problem.

I have a colleague who spent twenty years in the army before retraining as a soil scientist. 'It's interesting now to think back about the soil with what I know now, but at the time, it's always a problem. Dust or mud,' Cameron says.

He was in Indonesia helping out after the 2004 tsunami. Everyday routines, even things as simple as driving and walking, became harder. 'People were living in the mud,' he said. 'Like back to World War I, having to walk on duckboards. There were a few vehicles with communications equipment but it was too wet to drive. So they had to drive to the edge of the tarmac, but not much further as you'd get stuck.' Then they had to lay down boards to try and walk further. 'The whole place was covered in black, stinky mud from the wave. The stink not helped by the bodies strewn throughout.'

On another expedition, up at Shoal Bay in Queensland, there was an armoured personnel carrier trapped in mud: 'An armoured vehicle, weighing about eleven tonnes. It was carrying ten guys in the back when it went down. The vehicle was about 2.5 metres high,' he says. 'The whole thing sank except for the back corner.'

War. It's a big subject, so let's drill down. In this chapter I want to look at two specific aspects of soil in warfare: the

D-Day battle plans, in which information about the soil was so important that men were sent on covert soil-sampling missions; and what to look out for if you ever need to dig an escape tunnel.

D–Day

A soil auger left on a beach wouldn't normally warrant much more than a passing glance. Unless the beach was on the Normandy coast in France and the year was 1944. That single piece of metal could have undone nearly a year's planning for the D-Day battles.

World War II was more than four years in. Germany still controlled much of Western and Central Europe, though things were turning bad for them against the might of the Soviet Union on the Eastern Front. The Allies had defeated Germany and Italy in North Africa, and were fighting to gain a firmer foothold in Italy itself. The Allies wanted to invade German-occupied France and move through the continent, with an ultimate goal of attacking the Rhine-Ruhr region— Germany's industrial heartland.

It would not be easy. The Allies were up against Hitler's Atlantic wall, over 5000 kilometres of fortifications including mines and tank traps aimed to protect Europe's west coast from Allied invasion. The Dieppe Raid in August 1942 had showed what would happen if the Allies attacked the Germans without enough firepower or planning. In that instance, lack of naval and aerial support, along with difficult beach conditions, meant only twenty-nine out of fifty-eight tanks landed successfully. Of those twenty-nine, twelve got bogged at the sea edge. Fifteen made it over the sea wall but were then thwarted by the barricades. Within six hours the Allies had begun a retreat and within ten hours, over half of the six thousand men who landed had been killed, wounded or taken prisoner. It did not help that the Germans knew about Dieppe and were prepared for the attack.

The terrible outcome at Dieppe influenced the D-Day planning. As Vice Admiral Lord Louis Mountbatten said later, 'The battle of Normandy was won on the beaches of Dieppe. For every man who died in Dieppe, at least ten more must have been spared in Normandy in 1944.'

If the invasion was to have any hope of succeeding, the Allies needed a multi-pronged attack, deploying serious aerial and naval firepower to support Operation Neptune, the biggest amphibious assault in Western military history. The largest amphibious attack of World War I was at Gallipoli in 1915—where insufficient knowledge of the terrain (along with an unexpectedly fierce and well-prepared Turkish defence) led to disastrous results.

The Normandy point of attack, chosen with experience gleaned from Dieppe and other battles, was the northern French coastline. It was within range of aerial support from Allied fighter planes based in the UK; it was close enough to England to allow a swift crossing (although even then, due to terrible weather, for some the crossing ended up taking sixty hours) and close also to the major port of Cherbourg; and it had beaches rather than cliffs. Crucially, they were wide beaches, suitable to unload thousands of vehicles and soldiers and from which they could quickly move inland. At least theoretically.

There were four potential landing sites on the northern French coast. The Pas-de-Calais was ruled out as being too obvious. The crossing from Britain to France is only twenty-two miles at that point, making it the most logical place to attack—and indeed, the Germans had concentrated their forces there. Brittany was too far away. That left the Cotentin Peninsula or Calvados to the east.

With the Cotentin Peninsula initially chosen as the preferred landing site, British geologist W. B. R. King was brought in to assess how well the landscape could deal with the invasion. Taking geologists to war was still a relatively new idea, though Napoleon took at least four to help understand the local resources when he invaded Egypt in 1798. In World War I the British Army had just one geologist by 1915—King. Soil science was still a new field, and often kept firmly in the realm of agriculture (as it often is today). As World War I progressed, military geologists were tasked with finding quarrying supplies, deciding where to drill boreholes for water, guiding military tunnelling, quarrying for aggregates, and compiling 'going maps' that showed where tanks could travel across the landscape. Like much of soil science and geology even now, the work tended to be unsexy but critical.

Examining geological and topographical maps of the Cotentin Peninsula, King found the landscape hilly, eroded, full of hedges and ditches, and with waterlogging issues.

The mostly loam and clay soils would hold up all right if the weather was dry for a few weeks; if there was rain it would be a huge impediment. Machinery and people were likely to get bogged (something the Germans later took advantage of by flooding the areas to stop paratroopers landing). It would also be difficult to construct the temporary airfields needed to keep up aerial support. Temporary airfields could be quick to construct. The Saint-Pierre-du-Mont airfield, for example, was operational in just over one day, with essential work beginning on 7 June 1944 (D-Day+1) and completed by 6pm on 8 June. Without local airfields the planes would need to fly back and forth to England to refuel.

King noted that better airfield terrain lay in Calvados, to the east. The land was less undulating and its well-drained soils, overlaying limestone, were better to build on. According to chief engineer Major-General J. D. Inglis, King's appraisal of Normandy geology 'was, in fact, one of the main factors which led to the selection of the beaches eventually used'.

Once attention turned to Calvados the next step was to know with some certainty how the beaches would cope with a massive assault. Sand usually drains faster than clay but that doesn't mean you won't get stuck in it. Soft sand can be particularly problematic. Could the beaches support fifty thousand vehicles and over a hundred and thirty thousand troops? Or would they sink into the sand—sitting ducks for German fire?

The sandy beaches of Normandy were not likely to trap vehicles and men, like those clayey battlefields of the Western

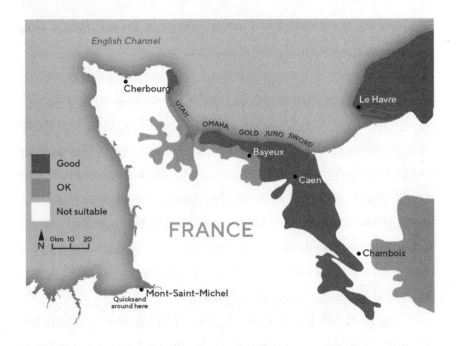

This map is a simplified version of the airfield construction maps from the Shotton Archive. The original map was made in 1943. Normandy was mapped using a grid, with soil type (rubble and sand, loam, clay, marsh etc.) on one side, and the number of potential airfield sites on the other (many, some, few, very few).

In my simplified version the 'good' areas have rubble/sand and loam soils, plus 'many' to 'some' sites. The 'OK' land has 'many' to 'some' sites on clay soil. All other areas have few or very few sites, or are on clay or marsh or waterlogged soil.

Front—or even Shoal Bay—had done. But they might very well stop tanks and severely slow the attack.

Engineers like sands that are well graded, meaning there is a relatively even distribution of different particle sizes. Imagine putting a squash ball, golf ball, tennis ball, baseball and volleyball into a big tub and giving it a good shake. The smaller balls settle into the gaps made by the larger ones, making a more stable surface—whereas if you only use tennis balls, the gaps remain. This type of sand is considered 'poorly graded' and does not pack together well; it is great for plants, as it means more pore spaces that can hold water and oxygen, but is less favourable for stability and for driving on.

So, on top of needing to understand the tides, weather, wave patterns and other nautical issues to make the journey across the Channel, the Allies needed as good an understanding as possible of the beaches themselves. Sailing across and collecting samples for compaction testing was out of the question. The details of the invasion (though the Germans anticipated it) needed to be a secret.

The Allies put together a team of scientists to tackle the task. The two that have made it to the forefront of the history books and who spearheaded the very creative research needed for the task are Fred Shotton and John (J. D.) Bernal. Shotton was King's protégé, a British geologist who had worked as a military geologist with the Middle East Command (and who had a keen interest in ice ages, beetles and fungi); Bernal was an Irish crystallographer, a Marxist who had earnt himself the nickname 'the Sage' for his polymath abilities.

Once they started researching, attention turned away from the sand and towards much more problematic material beneath the sand: clay and peat.

Under the cover of a fake photography competition, the BBC had put out a call to its listeners to send in photos of the French coast, which gave the team millions of photos to study. Bernal used old records from historical-society excursions that had recorded things like plant varieties and snail species, useful indicators of the nature of the local soil. An article from one of these excursions concerned a priest who walked out into the bay at low tide after a storm had washed away most of the sand to reveal a peat layer, and a Roman coin buried in it. The French Resistance smuggled four sets of geological maps out of Paris, including one created by the Romans that also showed large areas of peat (an important fuel resource). Bernal recalled visiting the beaches on a holiday and finding himself swimming alongside peat fragments.

The peat presented a problem—particularly since it seemed to be overlain by dynamic sand that washed in and out with the tides and storms, making it hard to predict in advance where the peat would be exposed and vehicles bogged. Storms in March 1943 changed the distribution of sand so much that the landing maps had to be redrawn.

Peat is highly organic, made from decaying plant matter. It forms when the vegetation can't decay as fast as it builds up, usually in waterlogged areas because there isn't enough oxygen for the microbes that break down the vegetation. Over time, layers of peat form, making a soft, wet, spongy medium that is excellent at trapping unsuspecting vehicles and people. Anecdotes abound: a farmer on South Australia's south coast who found himself in a tractor sunk up to the bottom of the cab, wheels completely immersed; an unfortunate hiker who became stuck up to his waist in a Yorkshire peat bog and was sinking quickly when rescuers arrived.

Patches of clay presented similar issues. The goal, then, was to find areas with enough sand on top of the peat and clay to act as a buffer to give vehicles traction, and to see how deep the sand compacted when driven on. The Allies wanted at least thirty-six centimetres of sand over the peat. (This amount was chosen because the British racing motorist Sir Malcolm Campbell, who held the land speed record in the 1920s and '30s and was considered an authority on the matter, said that was what they needed.)

The Allies ran air photograph missions disguised as bombing raids, at least once flying Fred Shotton, lying on the floor of the aircraft and peering through a specially fitted glass panel, as low as fifty feet (sixteen metres) to get a close look at the beaches. On these excursions Fred found himself assisted unwittingly by the Germans, who, knowing an assault could happen somewhere, had got local farmers to help move obstacles onto the beach. Fred could see how the sand compacted when the farmers' carts passed over it, providing another piece of the puzzle for the Allies.

The team found beaches in Britain with similar geology that they could use to run compaction testing. Brancaster Beach on the Norfolk coast was subjected to bombs and practice military operations to see how various vehicles could drive over the thin layer of shifting sand. But despite all the evidence and testing, inference was not good enough. Over a hundred thousand men were going to land on those beaches. The Allies needed to be as confident as possible of what the conditions underfoot would be. The team needed samples from the Normandy beaches themselves, data that could only be collected by actually setting foot on those beaches.

On New Year's Eve 1943, Major Logan Scott-Bowden of the Royal Engineers and Sergeant Bruce Ogden-Smith set out in torpedo boats to reconnoitre the area around Luc-sur-Mer at the eastern end of the Calvados coast. Scott-Bowden and Ogden-Smith were members of the Combined Operations Pilotage Parties (COPP) and had been trained at Brancaster Beach* in the art of soil sample collection. Late December is not a pleasant time to swim in the English Channel but the Allies were hoping that the Germans stationed along the coast would be indoors in the warmth, celebrating the end of 1943 with ample liquor, and avoiding the icy temperatures outside.

The pair went over the side into the freezing water 365 metres from the shore. It's not a very long way to swim in a pool—a leisurely ten minutes for a decent swimmer wearing a swimsuit. But this was open water with strong currents, and the equipment list comprised an eighteen-inch soil auger,

---

* Presumably. According to the UK's National Trust, the COPPs were trained at Brancaster, so we can infer that Ogden-Smith and Scott-Bowden were among them.

ten-inch long sampling tubes, rubber sleeves (actually condoms) to store the samples in, a bandolier designed to receive and hold the samples in the order taken, a beach gradient reel, wrist-watch in waterproof container, underwater writing tablet and Chinagraph pencil (works on most surfaces including stone and glass), compass, two waterproof torches, copper acetate fish scares, twenty-four-hour emergency rations, a pistol, ammunition, a knife and a brandy flask. In short, the swim was brutal.

Strong currents swept Scott-Bowden and Ogden-Smith about 730 metres east of their target and into the path of a lighthouse sweeping the beach every sixty-five seconds; they could hear the German revelry coming from the garrison. Crawling ashore and lying flat every time the lighthouse beam swept past, they pushed the augers into the sand at various points along the beach, noted the sample locations, examined an area of exposed peat, then began the difficult journey back to the boat—during which they lost an auger and a knife.

They believed they were out past the low-tide line and that the implements would stay buried in sand, and maybe they did. However, at some stage during the multiple beach-reconnaissance missions,* according to Fred Shotton, an auger appeared on one of the Normandy beaches. With the desperate emphasis on secrecy so present in everyone's minds,

---

* The COPPs conducted full surveys of the beaches over several weeks, and Scott-Bowden and Ogden-Smith were not the only stealth soil samplers—one reference says there were nine—but they are unnamed because of the long tail of wartime secrecy regulations. They may have included Nigel Clogstoun-Willmott, who reportedly founded the COPPs, and Donald McColville, whose wife didn't find out about his covert soil-sampling activities until the secrecy act ended.

this wayward auger caused a certain amount of nail-biting and gave rise to what was reportedly one of Shotton's favourite reminiscences from the time. There was much debate over leaving augers along the other beaches as decoys—but augers were in such short supply that in the end they relied on the efficacy of other misinformation operations to mislead the Germans.

And importantly, the sampling mission by Ogden-Smith and Scott-Bowden showed that in one area what appeared in aerial shots to be patches of treacherous peat were actually seaweed-covered rocks.

About five months later, from 6:30am on 6 June 1944, the Allies swept across the Channel and into France. The weather was foul. The ideal conditions for the invasion were: a full moon and low cloud cover (to help the soldiers see); a low tide to maximise the landing area; light winds and gentle seas (to minimise the chance of capsizing the boats). Instead, the sky was blotted with thick clouds that pelted rain onto the churning seas below. The weather was bad enough for the Nazi commanders to deem an attack unlikely—so unlikely that Field Marshal Erwin Rommel, handpicked by Hitler to ensure Germany's defence of the Normandy region, went home to wish his wife happy birthday and give her a pair of shoes.

The Allies landed and the battles began. It took nearly six weeks for the Allies to get beyond Caen, but Normandy was a turning point in the war.

—

On D-Day+1, with battle still very much underway, J. D. Bernal made the highly treacherous journey across the Channel. It took three boats to get to the French shore. The first boat was bombed and caught fire. The crew jumped ship to an American boat that made most of the laborious journey amidst bombings and bullets, but had to transfer to a French ship before being able to land. From Bernal's diaries:

We jumped out of it onto French soil at one of the precise spots that I had studied so often and whose history and geology were more familiar to me than any other place on Earth...It was an incredible scene: the confused row of stranded craft, the queue of vehicles coming off the beach, some broken-down or bombed, derelict tanks here or there.

Bernal wandered along the beaches and went to an area where he had predicted peat over clay would be a problem and was unsettled by what he found. 'In it [the clay and peat], like flies in amber, were stuck every kind of vehicle except the jeep: tanks, lorries and even DUKWs.'*

It turns out that although Bernal, Shotton and their planning team were not completely satisfied with their analysis before D-Day went ahead, the commanders deemed the intel good enough. Frustrated, Bernal wrote in his diary:

It is always the same way: I may be right, I may even know that I am right, but I am never sufficiently ruthless and effective to force other people to believe that I am right and to act accordingly. All this was so unnecessary: it could have

---

* A type of amphibious landing-craft.

been avoided. If people had not thought that my objections were just theoretical and statistical, and that they were practical people [who] need pay no attention to them.

Shotton too had experienced the battle between theory and practicality. We scientists always want more research and information, but at some point things need to happen. According to Fred Shotton's daughter, Anne Black, the Americans were sceptical of all the detailed research. But if you visit the Lapworth Museum of Geology in Birmingham, in the Fred Shotton Archive I believe there is a photo showing trucks bogged in the sand, captioned in Fred's handwriting: *I told you so.*

Visit the beaches today and they look like most other beaches. Sand, water, seashells; sunbathers, in the right weather. The sand is light grey—mostly quartz—some particles sharp, others worn smooth by tides and time. But if you look closely, under a microscope, there are traces of the battles hidden among the grains of sand. Shards of bone, brass, steel, iron; glass spheres made from sand melted by the heat of explosions. In the samples collected from Omaha Beach in 1988 by Earle McBride and M. Dane Picard, about four per cent of the material was shrapnel and detritus from that fateful day in June 1944.

Tips for digging an escape tunnel

Tunnels have been part of warfare for thousands of years. More than just a means of escape, tunnels have been used for guerrilla warfare, for smuggling, and to destroy fortifications by tunnelling beneath the enemy then placing explosions—or, pre-dynamite, tunnelling under the enemy's walls then setting fires to collapse the soil. In some wars, such as the Vietnam War, tunnels were not just a means of battle; they housed entire combatant communities, complete with kitchens, hospitals and recreational areas.

Tunnelling is a very dangerous activity. Never try to dig a tunnel unless death by another means appears to be imminent. However, if you do find yourself held captive in a place where tunnelling is a legitimate means of escape, understanding the soil and geology of your surroundings gives you a big head start. The success of an escape tunnel depends on many things, including the type of soil you are trying to dig through. If we look at hand-dug tunnels throughout history, as well as those dug by animals, there are some clues to what soil makes for better tunnelling: sandy, basically. Many historical war-time tunnels are in sandier soil that is easier to dig, with the bedrock and watertable far below the surface.* Wombats prefer sandier soil with tree roots and rocks that help stabilise the burrows. They avoid clay: it's harder to dig, and can get waterlogged, making an uncomfortable home.

Before we begin: *do not* start digging tunnels at home for fun. Tunnels are prone to collapse, trapping and killing those

---

* What most historical tunnels also have in common is that they were dug in the soil, by hand, and as much as possible in secret. Using dynamite to blast a path through rock is a pretty big giveaway a) that you are up to something and b) of your location.

inside, as well as destabilising buildings above. There are fatalities every year just from people digging for fun at the beach.

The following tips use examples from three famous wartime tunnelling operations—Stalag Luft III, Cu Chi and the Western Front. The Stalag Luft III tunnels were made famous by the book and the movie *The Great Escape*. Stalag Luft III was a prisoner of war camp run by the Luftwaffe in Poland that held captured Allied air force personnel. For the year leading up to March 1944, Royal Air Force Squadron Leader Roger Bushell led the Escape Committee, who dug three tunnels: Tom, Dick, and Harry. Though the movie makes it seem like it was only a handful of Americans who organised the escape, in fact over six hundred people were involved: stealing the necessary equipment and supplies, digging, and disposing of the tunnel soil. None of the escapees were American. There was also no motorbike escape over barbed-wire fences. Of the seventy-six escapees, seventy-three were recaptured. Fifty were murdered by the Gestapo and twenty-three were sent to other Nazi POW camps. Only two Norwegians and a Dutchman escaped.

The Cu Chi tunnels in Vietnam were a huge network of tunnels that ran from north-west of Saigon (now Ho Chi Minh City) almost to the Cambodian border. During the Vietnam War, in which the United States favoured aerial carpet bombing, the Viet Cong went underground, expanding tunnels originally dug in the 1940s. In time, particularly in heavily bombed areas, the tunnel networks grew to nearly

250 kilometres, housing entire villages. In the context of the guerrilla warfare that ultimately proved so successful, the tunnels were a place to rest and recover and prepare for night attacks. In 2009 I went to a touristy part of the Cu Chi tunnels and saw some of the booby traps set by the Viet Cong: trapdoors hidden on the forest floor often led to big metal spikes on rotating pieces of timber, designed to maim and ideally kill.

On the Western Front tunnelling was an extensive and ongoing aspect of military strategy. While trench warfare continued above ground, both sides tunnelled towards each other, aiming to place explosives under the enemy.

1. Check how deep you can dig

Step one—start digging down and be prepared to dig deep. Tunnels too close to the surface risk the top collapsing in. The tunnels at Cu Chi in Vietnam were between 150 centimetres and twenty metres deep, about nine metres at Stalag Luft III and about twelve metres deep on the Somme in World War I.

The purpose of this initial shaft is to check how deep the bedrock is, whether you are going to encounter the watertable or aquifer, and what soil layer options you have to tunnel through. Where the watertable is close to the surface the tunnel is at risk of flooding and limited in how deep it can go. On the Western Front, the watertable fluctuated up to nine metres, often flooding the tunnels and turning the surface soils to mud. You can pump out the water, but it's noisy. If water is running into your tunnels and you are only one metre deep, you will likely need an escape plan B.

2. Decide which soil layer(s) to tunnel through

Look at the wall of soil layers in front of you and check the texture of each layer. You want to know whether it's more like sand, clay, or something in between so take a handful of soil and mix it with water, continually working the soil with your hand to try to make a bolus.* Straight sand won't stay in a ball, clay will be sticky and probably stain your hand, a loam might feel easy to work into a ball and a little spongy. Soil scientists usually work with at least twelve soil textures, with variations including clay loam, silty clay loam, sandy and so on, but for this exercise a rough guide to texture is enough. Whether a soil is sandy, clayey or something in between makes a huge difference to timing, effort and ease of secrecy. Sandier soils are usually easier to dig, but the lack of clay to act as a binding agent means the walls are more likely to collapse and will need structural support.

The Tom, Dick and Harry tunnels in Stalag Luft III were dug on sandy soils called Podosols. This soil had about ten to twenty centimetres of fine, dark brownish-grey sand overlying at least ten metres of yellow sand. The sand was very easy to dig but needed lots of structural support. When the Germans did a camp inventory after the famous escape they found four thousand bed boards, fifty-two twenty-man tables, thirty-four chairs and seventy-six benches had been stolen, dismantled and used to construct tunnel supports and ladders.

---

* Soil-science–speak for a small ball of soil made when working it by hand. We make boluses to determine the soil texture.

Clay is often harder to dig though, especially when the soil is dry, but it might be the best layer. Around Ypres on the Western Front the soil landscape was more complicated than at Stalag Luft III. Ypres has varying layers including sand, sandy clay, Kemmel sands, Panesilian clay ('bastard blue clay'), and Ypres clay (now called Kortrijk Formation). The Ypres clay, which is blue-grey when wet but oxidises to a dull brown colour, is a lot like the clay the London Underground goes through. It was the best choice to tunnel through as it was largely impermeable to water: any that seeped into the ground sat on top of this layer, staying in the layers above. Getting to it, however, meant digging metres and metres down through the other layers.

The Kemmel sands and aquifer were particularly problematic, being nearly always wet, and were nicknamed *Schwimmensande* (swimming sands) by the Germans. When dug into they spewed yellow sand into the shaft, making tunnelling nearly impossible if you couldn't stem the tide of flowing sand. Where this layer was thin, about less than one metre, it was manageable. But on higher ground the Kemmel sands were up to ten metres deep. Getting through the layer was a big enough problem that the tunnellers gave up in a few places.

If you're digging through clay, mineralogy is important. The Ypres clay contains a lot of the clay mineral montmorillonite, which tends to swell when wet. When it rained, the tunnel walls swelled so much that they snapped some of the support beams. The Panesilian clay above had more sand in it and was less prone to swelling, but was also prone to waterlogging, requiring pumps to remove the water.

As for digging technique, if clay is all you have to work with, consider clay-kicking, a method developed in the UK and used extensively on the Western Front (and to dig parts of the London Underground). It was known colloquially as 'working on the cross' because workers lay on a wooden cross that was wedged into the tunnel so they were lying at an angle of about forty-five degrees. Using a sharp-pointed spade with footrests on either side, tunnellers could dig about four metres per shift, compared to just under two metres when digging in the normal way.

The Cu Chi tunnels in Vietnam had the perfect soil for tunnelling. Situated on old alluvial soils, they were mostly clay and silt with some fine sand. They also had high levels of iron. Over time the rains leached the iron from the upper layers to accumulate deeper in the soil. When it dried, the soil set like concrete. Kaolinite was the main clay mineral. It's not prone to shrinking and swelling, meaning it was less likely to collapse. The watertable was deep, about ten to twenty metres down, and the roots of the forest above helped stabilise the soil.

These tunnels were so strong they could withstand sixteen-kilogram explosive charges without collapsing. In the open, you would need to stand about 340 metres away from a blast like that to be at a 'safe distance'. In the tunnels, the blast would follow the path of least resistance through the network, devastating the people and infrastructure in the tunnels but doing little damage to the tunnels themselves.

3. Dig when the soil is damp

This is not just for ease of digging but for noise too. The chalky soils towards the south of the Western Front were soft and comparatively easy to dig when wet, but set quite hard and stable when dry. This was useful from a structural perspective but problematic in terms of secrecy; if the soil was dry the tools clanged against the tunnel walls.

At Flanders, shallower tunnels were dug towards German lines to act as listening posts. Soldiers would crouch at the end of the tunnels and if they heard the Germans tunnelling, they would try to pre-empt an attack by digging towards the German tunnels as quietly as possible.

The Cu Chi tunnels were dug during monsoon season. Trying to dig in the dry would have been impossible.

4. Have a soil-dispersal plan

You are going to be moving tonnes and tonnes of soil—it needs to go somewhere. A soil-dispersal plan considers the best way to hide the excavated soil. In Vietnam, the Viet Cong put the soil into bomb craters they then farmed, or poured it into the Saigon River at night, or used the soil to build combat trenches.

At Stalag Luft III, the bright yellow sand particles were a completely different colour and texture from the dusty grey soil above, and much harder to hide. The disposal from the trouser-leg technique shown in the movie was the main dispersion method: prisoners would attach small pouches inside their trouser legs. The pouches were closed with a pin attached to a string which fed up through the trouser leg and into the pocket. As the prisoner walked around outside they would pull the string, releasing the pin, and the sand would fall out the bottom of the trouser leg. Another method was to stuff their socks with sand and then tip it out into the small gardens they were allowed to tend, raking the excavated sand into the soil. The dispersal organisation also hid soil in another tunnel that had been started but abandoned. By the time the Tom tunnel was discovered, prisoners had moved 166 tonnes of sand.

On the Western Front, both sides were tunnelling and trying to keep the locations a secret. The Allied tunnellers had trouble removing the bags of soil quickly enough so the Germans wouldn't see them. Captain Martin Greener of the 175th Tunnelling Company recounted to historian Lyn MacDonald in her book *They Called It Passchendaele*: 'If you collected a pile of sandbags with blue clay you had to get rid of it quickly, because the Germans would shell it just to see what was in it.'

5. Get captured with a soil scientist and miner

If you are going to get captured, having certain skills will make your escape tunnel efforts easier. A well-rounded crew will give the best chance of escape. Ideally, you want a soil scientist (a geologist will also do), a miner, an engineer and an electrician.

Other uses of soil in war

Here are two other tales about soil and war. The first is about Alfred Nobel, commonly known as the founder of the Nobel Prize, less commonly known as the inventor of dynamite.

Nitroglycerine, once called 'explosive oil', was synthesised in 1846 by Italian chemist Ascanio Sobrero, who noted that the oily, colourless liquid was highly explosive. It was also temperamental, sensitive to knocks and bumps and, when first developed, unpredictable: it was difficult to know when it would explode. Over time it degrades, making it even more volatile. It is unwise to go putting lots of nitroglycerine on trucks and bumping them all over the country.

After his brother was killed at their father's factory while experimenting with nitroglycerine, Alfred Nobel worked to make the product more stable and safer to handle. After a few false starts mixing nitroglycerine with sawdust, coal and cement, Alfred eventually tried kieselguhr, or diatomaceous earth, collected from a river near the family factory in Hamburg, Germany.

Diatomaceous earth is made from fossilised algae called diatoms. Deposits are mined throughout the world for a multitude of industrial uses including toothpaste, as an anti-caking agent, and in specialised potting mixes. When used in a growing medium, diatomaceous earth has the wonderful ability to hold lots of water and nutrients without getting too waterlogged.

By mixing nitroglycerine with diatomaceous earth and a bit of sodium carbonate, Nobel invented dynamite and the world was enabled to focus on exploding the landscape and each other rather than accidentally exploding themselves.

The second story regards the great bleached sands of Botany Bay in Sydney. At the outbreak of World War II, Australia found itself very short of optical munitions—glass lenses made from pure and optically precise glass. These lenses are used in rangefinders for artillery, gun sights on aircraft and periscopes on submarines. When the war began, Australia was still importing optical munitions from Britain. Supplies soon ceased and there was a scramble to make our own.

We had few experts who could grind the lenses to the required accuracy, but a worse problem was not having the glass to grind. Glass, in its simplest form, is made by heating sand until it melts—at about 1700 degrees Celsius. More 'pure' sand is mostly silicon dioxide, which makes clearer glass. Typically, Australian sand had traces of iron and manganese that clouded the glass so it could not be used to make lenses for telescopes and gun sights (but was good for brown beer bottles). The UK and the USA, upon whom we had been relying for such raw material, put a ban on their exports when hostilities broke out.

Yet after only six months of manufacturing, we had perfected the process and Australia was able to not only meet our defence needs but help out other nations as well. We produced 68,000 dial sights, 46,000 anti-tank gun sights, 10,000 tank periscopes and thousands of other necessary optical munitions, while supplying the USA with 57,000 pounds (26,000 kilos) of optical glass and New Zealand with 4700 pounds (2100 kilos) of glass.

The company that did most of the work (with significant help from university physics and chemistry departments, and possibly agricultural soil departments) was Australian Consolidated Industries (ACI), the local subsidiary of Imperial Chemical Industries (ICI), which ran a glass factory in Waterloo. Botany Bay was on their doorstep, and there, beneath an organic A horizon, lay metres of pure, bleached white sand, leached of most minerals and impurities. Sand sufficiently depleted of almost every other element to make clear glass and lenses: optical munitions for the war effort.

Quicksand

THIS MAY BE a niche opinion, but quicksand gets its own chapter just because it's great. Many of my favourite childhood movies featured quicksand in some form—I watched *The Princess Bride*, *The Neverending Story* and *Jumanji* (where the floor becomes quicksand) until the VHS tapes wore out. If I'd been born twenty-five years earlier I would have had even more quicksand to appreciate. A movie buff has estimated that one in every thirty-five movies (about three per cent) in the 1960s used quicksand to generate an element of suspense. Mostly the characters are in either a swampy forest or a desert when they step unawares onto the quicksand and become stuck. The more they fight it, the faster they sink. Only a conveniently located branch or vine and a strong set of arms, or sometimes a faithful horse, saves them from being pulled under to their doom.

It's not just a movie gimmick, though; quicksand is very real. It's found in tidal areas—near beaches, rivers or marshes—and it really does trap the occasional person or animal. There is quicksand near a salt lake in Iran that is rumoured to have swallowed camels whole.

But how accurate are the tales and movies? Does stepping in quicksand really mean certain death unless there is a convenient vine or muscular associate close by? Will you sink faster the more you struggle? How do you escape if you find yourself trapped? And why do our protagonists always step into the sand unawares—are there no clues?

Alas, for the latter question, no: there are very few visual clues. Part of what makes quicksand so treacherous is that it is very difficult to recognise by sight. Areas where it tends to occur might be signposted, but it is difficult to tell exactly where the dangerous stuff is. This is because the way in which quicksand forms, and what it is made from, leave the surface looking solid and safe while there are interesting phenomena happening beneath.

Despite the name, quicksand is more than just sand. It is usually a mixture of about forty per cent sand, five to ten per cent clay, and salt water. How this mix becomes so treacherous can be explained by imagining a bucket of apples with bits of toffee sticking the apples together where they touch. The apples are the sand particles and the toffee is the clay. On most soils, if it rains the water will filter through the pores (the gaps between the apples) with little change to the structure or stability of the soil. You can walk quite happily along the top without fear of sinking, because the particles are packed together firmly enough to make a stable surface. On a clayey soil you might sink a little but will be able to pull your foot out if need be.

Because quicksand is tidal, water seeps in from the sides or below. Salt water has a greater tendency than fresh water to weaken the clay that is holding the sand particles together, so in time the sand particles are suspended like a very fragile house of cards: a very loosely packed sand held together just enough by the clay that it doesn't collapse under its own weight.

From above, quicksand looks like a firm and mostly dry surface. Two clues that you might be on or near quicksand are if you are walking in a tidal area and your footsteps immediately fill up with water, or if you see water bubbling up from below (but this is rare).

The recipe for quicksand is delicate. Quicksand with almost no clay does not last long because it naturally compacts under gravity. This sort of quicksand forms in the Netherlands when making new polders, areas of low-lying land reclaimed from a body of water. Patches of quicksand in polders last for a few months. But add five to ten per cent swelling clay, and the sand can be sufficiently bound together to persist for years.

The sand particles cannot be too big for the same reason: they will pack together under gravity. A sample of Iranian quicksand, the very stuff reported to have swallowed camels, had a typical grain size of 60 micrometres (0.06 millimetres) which is classified as a 'very fine sand', nearly the smallest size a particle can be before it is considered silt.*

Together the sand, clay and salt water become a shear-thinning non-Newtonian fluid—a mix that doesn't adhere to the regular rules for liquids. Quicksand is solid when at rest but liquefies when agitated (like toothpaste). Agitate the quicksand—say, by walking on it—and it stops acting like a solid and starts behaving like a liquid (liquefaction): you sink.

There is an excellent video on YouTube by a man in Northern Arizona showing a patch of quicksand that needs stress—in the form of him jumping—to turn from solid to liquid. Skip forward to the three-minute mark. The more he jumps the more spongy the sand becomes, until it looks like 'Jell-O', as he describes it. Videos of people liquefying the soil around Mont-Saint-Michel in northern France show a similar thing, though it makes me anxious to watch them. The tide comes in at about six kilometres per hour—not too stressful—but rises an average of ten metres. You want to be confident you can get out of the quicksand you're playing in before the tide starts to come in.

This soil liquefaction is also what makes buildings sink during earthquakes. The shaking shifts the soil particles around so much that the friction between the particles is lost and the soil acts like a liquid and the buildings sink. It behaves

---

* Smaller than 0.02 millimetres and it is silt; smaller than 0.002 millimetres and it is clay.

the opposite way to Oobleck,* a mix of cornstarch and water which is liquid when at rest but solid under force. Stand on a bucket of Oobleck and you'll sink in; run on the spot and you won't.

You get trapped in quicksand because as you sink your legs displace the water from between the particles, so the sand starts to set hard again—it becomes a solid with you stuck in the middle. The idea that the more you struggle the faster you sink is true, at least to start with. The more you move, the more you agitate the sand; the more it liquefies, the more you sink. The good news—'good' being relative—is that you probably won't sink completely. Quicksand has a density of about two grams per millilitre and you are about one gram per millilitre: you're not dense enough to go all the way under. You're likely to sink to about your waist—your chest at most—before the sand settles around you, setting hard so you can't sink any further.[†]

This is still very problematic. If you are stuck up to your chest you might suffocate from the compression, as you can't expand your lungs. The pressure could also cut off blood flow. The most likely fatal quicksand scenario, though, is that the tide will come in and you'll drown.

National Geographic has a great video in which the presenter willingly gets trapped in quicksand to see what it's like and how to escape. It takes him eight stressful minutes to sink

---

* The name is from the Dr Seuss book *Bartholomew and the Oobleck*, about a sticky substance that is threatening the Kingdom.
† Which calls into question the credibility of the Iranian camel-eating quicksand, I suppose. Perhaps camels are denser than people.

to his waist, whereupon he becomes very anxious—saying it feels like he's stuck in concrete—despite the presence of a fully equipped rescue team close by.

How to escape

The trick to escaping quicksand is to get water back in between the sand particles that have settled around your legs. At this point you need to move. In the National Geographic video the rescue team stuck a perforated metal hose into the sand near the presenter's legs to push water back into the area, digging around at the same time. This reduced the force around his legs and allowed them to pull him out.

The strength required to escape quicksand if you can't get water in is phenomenal. Experiments on the Iranian camel-eating quicksand using a model foot required a force of a hundred thousand Newtons—about what it would take to shift a medium-sized car—to pull the foot out. As the researchers wrote in a later paper: 'The classical Hollywood scene of the hero being pulled out of the quicksand by his horse…is very wrong: the horse is likely to pull our hero into two pieces.'

If you don't have a convenient rescue team handy, the advice is to wiggle your legs to allow water to flow back into the sand, and lean back or try to lie down to distribute your weight evenly across the sand. Also yell for help as loud as you can. The best approach is to not get stuck in the first place.

—

Many movies, including my favourite of all time, *The Princess Bride*, feature dry quicksand, which is, as the name suggests, like quicksand but without water. In *The Princess Bride* Wesley and Princess Buttercup are walking through the fire swamp when a patch of lightning sand swallows the princess whole. Once Wesley realises what has happened he grabs a vine from a nearby tree and dives in to save her. Dry quicksand also appears in *Lawrence of Arabia*, *Miss Fisher and the Crypt of Tears* and *Indiana Jones and the Kingdom of the Crystal Skull*.

Dry quicksand is not known to science—nor indeed outside of folklore and popular culture. None has been found in nature. But it has been created in the lab by blowing air through sand. The idea is similar to regular quicksand, where the sand particles are so loosely put together they can only just hold their own weight. Put weight on top, and the weight sinks in, causing a jet of sand particles to shoot into the air as the sand collapses and buries the object. Or princess, as the case might be.

Memory

Chemical memory

'After scraping it away carefully with a trowel, most archaeologists take the soil, throw it into a two-millimetre sieve and discard anything that falls out the bottom,' Roy Lawrie says with a shake of his head.

For an archaeologist in Western Europe this might be a worthwhile exercise. After all, you can barely plant tomatoes without hitting a pottery shard or old coin. Take, for example, London's Crossrail project, a seventy-three-mile (117-kilometre) underground crossing connecting London's east to its west. Since construction began in 2009, archaeologists have found, on average, something of significance at least twice a day. A mammoth jawbone, a skeleton from a Great Plague victim, a multitude of skulls from Roman to more recent, fifty-five-million-year-old amber and part of a small barge or fishing vessel from about the 1200s are just some of the ten thousand-plus finds.

Although Australia has one of the longest histories of occupation, it's much harder to find evidence of human activity before 1788. Indigenous Australians were remarkably low impact: the landscape was the significant feature, and building structures wasn't particularly high on the to-do list. To find our nation's history from an archaeological point of view, you need to look smaller. You need to look chemically—at the stuff that falls out the bottom of the sieve.

This is where Roy specialises. Retired now, though still happily digging holes, Roy lived a double life as a soil scientist and archaeologist: mapping soils and landscapes for the Department of Agriculture and helping trace history when archaeologists needed a different perspective. The archaeology part of his career came about by accident. One fateful day in 1980, Roy glanced out the window from his office in Parramatta and saw someone digging a hole in the carpark opposite. Within minutes Roy was peering into said hole. At the bottom was archaeologist Ted Higginbotham, busy excavating a brick barrel drain built by convicts. Roy and Ted got chatting and thus began a new branch of Roy's career.

Roy and his wife Lisa have graciously travelled up from Wollongong to meet me at the Museum of Sydney, the site of Roy's first big archaeology job. As a retired chemist and environmental scientist, Lisa often assists with site investigations.

'How would you tell if a fire in 1815 was started by an

Indigenous Australian or a European settler?' Roy asks with a grin. 'I'll give you a clue: it's to do with fuel.'

He leans in, watching my face closely. I am touched that he thinks I know the answer, but mildly panicked, since the only knowledge I have comes from a high-school history class in which a visiting Aboriginal man explained smoking ceremonies to us. To the delight of every teenager in the room, he started a fire in the pot plants on the school hall stage; to our collective disappointment, the school did not burn down. But this is no help with my current pop quiz.

'The heavy metals were through the roof!' Roy hints.

'...Zinc?'

'Yeeeeeess!' Roy sits back and winks. He has a way of drawing out certain words, mostly 'yes' and 'right', that make you feel like you're in on some special secret.

Zinc is an essential plant nutrient that tends to concentrate in leaves: there's usually more zinc in the leaves than in bark or branches from the same plant. It was more common for Indigenous Australians to start fires with leaves and bark. The Europeans, by contrast, tended to cut kindling and avoided the smoky leaves.

As fire burns, the leaves and branches are broken down into ash, and the nutrients released into the soil. This is partly why slash-and-burn agriculture (cut down the forest, burn it to fertilise the soil, then plant crops) is more common in regions with limited access to fertiliser—it's a free source of nutrients.

It also means that if you test the soil beneath an old

fire pit, you might be able to work out what was burning. In 2016, when Parramatta in Western Sydney was getting a facelift, archaeologists found three remnant fireplaces near the old library. Roy was called in to investigate the hardened orange lumps of burnt clay: were they significant fire sites of the Burramattagal people?

The burnt clay chunks were localised—so it wasn't a big bushfire that burned the soil—and they were fifteen to twenty centimetres deep. Soil is a poor conductor of heat; until all the water in the soil evaporates, it stays under one hundred degrees Celsius, and it would have taken many hours for the heat beneath a small fire to start baking the clay subsoil, meaning these fires were most likely deliberately lit and tended.

Second, levels of twentieth-century heavy metal contaminants, particularly copper and lead, were low, with the exception of the aforementioned zinc. These fires burned before modern industrial times.

Combining the soil analysis with other historical and archaeological evidence, the archaeologists concluded that the old library area could have been a meeting place of the Darug Burramatta clan, where smoking ceremonies were probably undertaken to welcome visitors to the site.

It is because of chemical evidence like this that Roy, Lisa and I are sitting in a café next to a museum, not on the ground floor of an office tower.

In 1983, the government was negotiating a development on the corner of Bridge and Phillip streets in Sydney. A developer wanted to build office blocks on the site, but archaeologists had uncovered the original building footings, mortar and even some pottery shards from Australia's first Government House.*

With the Bicentennial coming up in 1988, the topic of archaeological finds from the start of the colony was a hot one, and the plan to develop the site became controversial. The developer wanted to know if they could build on the undisturbed parts next to the old building footing. If there wasn't anything historically significant there, why not just fire up the cranes and dozers? Roy was called in to put a boundary—horizontal and vertical—around the site. How far had the activities around the old Government House extended? What was the extent of human impact on the soil?

Further than first thought, as it turned out. Although the soil looked physically untouched on much of the site, the chemistry told a different story.

'The first time I saw those [soil] pHs, I told my offsider, a young student from the University of Sydney, *you've fucked up, go do them again*. He did, and got the same results. It wasn't a mistake, it was a discovery,' Roy says.

Sydney has very acidic soil. A pH of 4 would not be surprising—intolerable for many veggies, but certainly suitable for the native vegetation (also blueberries). But the samples

---

* If you ever visit the Museum of Sydney, take care to note the pavers out the front. The large open space is predominantly dark grey, with occasional beige pavers. These beige pavers mark the footings of the original Government House. The Museum stands behind the pavers, where the gardens once were, and houses the artefacts found on site.

from the areas where the developer wanted to build had a pH of up to 9. Just over double what you'd expect? No: pH is logarithmic. It was a hundred thousand times higher. On top of that, the levels of calcium and phosphorus were eight times higher than expected. This was not untouched soil. Something had been done to it that had significantly altered its chemical profile.

Adding lime is the most common way to raise pH in agricultural or garden soil, but anything with carbonates that can neutralise acidity will work. Concrete dust, crushed-up shells, even chalk can make a soil less acidic. This was early 1800s Australia—there wasn't much concrete going around, and we don't have any chalky geology nearby. Maybe limewash from the walls had somehow found its way into the soil? But to get that much of a pH change would take about fifteen tonnes of lime per hectare of soil—more than you would be likely to get from limewash runoff.

Roy eventually found the answer while rummaging around in papers from the Forestry Corporation, one of which described the mineral content, including levels of phosphorus and calcium, in different types of timber. Ash from *Angophora* species is nearly sixty-three per cent calcium. Fairly useless as a building material (these gums tend to crack if you look at them sideways), they were an abundant source of firewood, and the ash was used to make plaster. In Peter Cunningham's 1827 guidebook for settlers, *Two Years in New South Wales*, he describes how to build huts covered with plaster 'whitewashed with lime, plaster of Paris, or apple-tree ashes and sour milk'. The smooth-barked apple tree is the local name for *Angophora costata*, which is common around the Sydney region. It is very likely the same species referred to in 1827 and tested much later in the twentieth century.

The first Government House had many fireplaces, and the outbuildings for washing and cooking would have demanded fuel year round. When the fireplaces were emptied, the ashes were dumped outside. Over time those ashes worked their way into the soil, neutralising the acidity, raising the soil pH—and marking the history for Roy to dig up 160 years later.

You can find the same chemical anomaly in Hyde Park in Sydney, if you're motivated enough to dig down a metre and a half to find the original topsoil. Development has a tendency to smooth out the landscape—cutting the tops off the hills and tipping them into the low bits. In the 1920s Hyde Park was dug up to build St James and Museum stations (and the adjoining tunnels), part of the city circle train lines that run underneath the park. They were covered up again with some of the excavated soil and rock. Now a beautiful park with grand buttressed figs and a delightful café for a cheeky lunch wine, part of Hyde Park was once a dumping ground for rubbish that, prior to modern methods of heating, would have included quite a bit of fireplace ash. In the nineteenth century Hyde Park was flanked by houses and other colonial buildings, all with fireplaces.

Footprints

A hunter picks his way through sparse shrubs and grass, shivering as a cold wind blows in from the south.

In the distance, a flicker of movement.

He breaks into a jog, clay squelching between his toes even though it hasn't rained for a while. The grasses rustle underfoot, but not much. Even at this speed, he knows where to step to keep speed and stay quiet, avoiding the small puddles along the edge of the lake. His footprints fall alongside fresh tracks: elongated rectangles with three claws.

As he weighs his spear and plans the throw, the creature pops its head up, alerted somehow, and he starts to sprint. His long legs cover two metres of sticky ground with each stride—at thirty-seven kilometres per hour, he could keep up with Usain Bolt in a race.

The creature jerks its body to stand front on and crouches: the slight downward hunch before the spring up and away. Still sprinting flat out, the hunter raises his spear and the weapon flies, slicing a perfect arc through the air.

Twenty thousand years later, scientists peer down at the hunter's tracks among the confusion of footprints now set in stone. Hot, dry air blasts across the sand dunes, fluttering the plastic tape that marks off the site—the same wind that shifted the dunes to expose the rock in the first place.*

---

* Soil, particularly sand dunes, has a tendency to shift about, revealing treasures that lie beneath. In the winter of 1997, powerful storms eroded the Tillamook Coast in Oregon, exposing an ancient Sitka spruce forest at

It's 2003, and this is one of the most exciting historical landscapes in Australia. If footprints are your thing, it's one of the most exciting landscapes in the world. Here, at Mungo National Park in New South Wales, 315 kilometres south-east of Broken Hill as the crow flies, more fossilised footprints have been found than anywhere else on the planet.

Footprints might seem on the lower end of exciting fossil finds: not exactly an undisturbed brontosaurus skeleton. But because not all creatures were considerate enough to die in the right place at the right time and in the right conditions to become fossils, experts have learnt to look for less obvious evidence of living history. Trace fossils like footprints, burrows, borings and old root channels are the circumstantial evidence that life was here. Burrows or borings can show what insects or worms were wriggling about; tooth marks allow us to deduce who was chomping on what.

My favourite are coprolites: fossilised excrement. Not only can we tell how a creature walked, hunted and ate, we also have clues regarding their bathroom habits. Take, for example, the dinosaur dump specimen from Canada—forty-four centimetres by sixteen and filled with bones—that scientist Karen Chin concluded had been excreted by a T-rex which had dined on other dinosaurs. Or the discovery by George Poinar and A. J. Boucot in 2006 (while they were literally polishing a turd; it had gathered a little dust since its last showing in 1903) of protozoan cysts and nematode eggs indicating the

---

the edge of the ocean. The Neskowin ghost forest tree stumps poke out of the sand at low tide.

presence of intestinal parasites. In York, England, there's a shiny Viking poo on display, proudly named The Lloyds Bank coprolite. I do wonder if marketing put much thought into that one, or if some now-ex-employee got the last laugh.*

The Mungo† prints were discovered in 2003, when park ranger Mary Pappin Junior walked into the long, dry basin, glanced down and noticed a single footprint etched into the claypan. Researchers quickly mobilised, measuring and cataloguing the echo of the community that walked through thousands of years ago.

Finding the footprints wasn't as simple as putting eager undergraduates to work with dusting brushes. Because the prints were embedded in various layers in the soil, excavating deeper prints would mean destroying shallower ones. Ground Penetrating Radar (GPR), a tool that lets you peek underground without the need to dig, revealed over seven hundred prints made by at least twenty-six different people and some marsupials, embedded in eight layers of compressed clay. GPR is a popular tool in archaeology (and construction and the military) and has been used to find terracotta warriors, a Viking ship buried near a vicarage in Uppsala, Sweden, and even the remains of King Henry I.

A small GPR looks a bit like a push lawnmower, but

---

* In 2003 someone dropped it and broke it while showing it to visitors.
† This was not the first time Mungo caused an international stir. In the 1960s, just a few kilometres from where the footprints were found, the bones of a woman were (also accidentally) found. Analysis showed the 'Mungo Lady', as she is now called, is the oldest evidence of ritual cremation in the world.

instead of cutting grass it shoots radio waves into the ground. Objects and soil layers reflect the waves back to a receiver. The strength of the signal and the time it takes to come back can give clues to how deep the object is.* Soil clay content, salt content and moisture content affect readings.

The age of the Mungo prints was established by measuring the amount of radiation stored in the soil. All soils contain trace amounts of radioactive isotopes of certain elements, notably potassium, uranium and thorium. As these isotopes decay, they produce radiation (a luminescence signal) that is absorbed by some soil minerals, such as quartz or feldspar; the luminescence signal accumulates in the dark underground. Not to any extreme amounts (fear not for attack of the giant earthworm), but enough that we can measure what's there. That measurement gives an indication of how old it is. Once exposed to light, the radiation dissipates—you can't just dig up a sample to test. The solution is to measure radiation by firing blue or green light at the soil while it's undisturbed.† The quartz grains around the Mungo footprints hadn't seen light since they were buried by sand blown in about twenty thousand years ago.

After months of careful work researchers could see who had walked where, and had a rough idea of when. They measured print and stride length to estimate age, height and speed, which is how we know one hunter may have run at thirty-seven

---

* This is a very simplified description of how GPR works.
† Known as optically stimulated luminescence.

kilometres per hour.* Prints that start closer together then become further apart indicate running or picking up speed. The various sized prints indicated a group of men, women and children.

Though not tried at Mungo, some scientists use footprint depth to estimate weight, a tricky exercise as this method depends on soil clay content and moisture content. As any enthusiastic four-wheel driver knows, the wetter the clay gets, the deeper you sink,† which means a very wet soil can overestimate weight.

But to truly see the nuance in the Mungo tracks, researchers needed expertise from those trained in the art. Consultant trackers, Pintupi people from Central Australia, were brought in, and they pointed out minor movements and markings that brought the prints vividly to life:

About twenty thousand years ago a group of people walked along the edge of a wetland, in a clay soil soft enough to capture footprints. Children scampered about and drew in the mud. One child moved in the way children do: meandering in one direction then suddenly another, all the while changing pace, walking, stopping, running. A woman shifted her baby from her right hip to her left and back again, a one-legged man hopped across the expanse, someone else dragged

---

* 'May have' because there are only a few prints to measure by, and when you run there are explosive bursts that are faster than your average running speed.
† Sometimes this is not a bad thing. A colleague of mine had a one-tonne excavator bucket dropped on his foot. If he hadn't been standing on soft clay after a week of rain, it would have crushed his foot but instead it just pushed his steel-capped boot into the ground.

a dead animal along the shore. At some stage eight little emu chicks, about two months old, walked through. And there are skids across the ground where the hunter's spear missed.

It was the height of the last ice age, with maximum ice sheet coverage around the world. Icy winds blew off glaciers in the south while sand dunes blew in from the west. The climate was cold—about five to ten degrees colder than modern temperatures—drying, and changing. Sea levels were falling, deserts expanding and forests receding. In another fifteen hundred years the lakes would have dried up. There may still have been some short-faced giant kangaroos hopping about, but populations of the modern kangaroo were increasing.

Had the group walked across dry ground further away from the wetland, we wouldn't have these tracks. And, of course, those who walked through at different times left no marks at all.

How to become a fossil

If leaving a footprint that lasts twenty thousand years is a chancy and unlikely business, becoming a fossil is a tricky endeavour too, particularly for land dwellers. I can't find any numbers to say exactly how rare, but you're more likely to become a billionaire than a fossil.

You need to die somewhere watery—a river bank or muddy shoreline (being buried under layers of volcanic ash or lava might also work). Even though palaeontologists often seem to be in hot, dry, rocky places, that's only because it's easier to find fossils there. To find fossils you need to be where the rocks are and you've got more chance of stumbling across one in a rocky desert than a forest.

The type of soil you die in or near is important. Coarse sand—beach sand—has too much oxygen in the pore spaces between the particles. Think of how quickly water soaks into sand compared with how it pools on top of clay. That water is displacing oxygen. Too much oxygen encourages soil microbes to decompose your body too fast. If sand is where you want to take your last breath, you're better off going to a desert and hoping you'll mummify. For fossilisation, soils with smaller particles like silt and clay are a better choice— like mangrove swamps or marine mud. They're also better for detail. The smaller the particles, the better the impression. Compare a footprint in coarse sand to a footprint in clay; the detail is significantly better in the clay.

Next you need this silty mud to wash over and cover you very quickly, before something drags you off and eats you. To have even a hope of becoming fossilised, you need many layers of sediment to build up on your corpse. The best way for this to happen is for the sediment to wash in, so the waterbody you've chosen needs to move. A river is good; a pond where nothing covers you up is not. The sea floor is a good spot to become a fossil, which is why most fossils are sea creatures. However, chances are you won't be discovered again—and what's the point of becoming a fossil if you don't end up on display in a museum?

If you're really dedicated to being preserved for posterity, it's best to be buried on a floodplain, just before a flood that will wash in lots of sediment to bury you even deeper. Over time, more and more sediment will wash in, adding layer after layer. The finer particles settle out, moving to the floor. The weight, building and building, compresses the soil and encourages it to start turning back into rock. This is why most fossils are found in sedimentary rock, formed by washed-in layers of sediment. Generally, metamorphic and igneous rocks are subject to too much heat and pressure to preserve fossils. It's not impossible, but it's rare. Sedimentary rocks form under the gentler pressures and temperatures that may preserve life forms.

Eventually the soft parts of your body will decompose, leaving only the hard stuff—bones and teeth. If you've timed it right, this is where the switch happens. Your bones and teeth degrade, leaving gaps that will be filled with something else, such as minerals that crystallise: it's like making a cast.*

Now you're a fossil: buried safely in sedimentary rock, waiting to be revealed by erosion and rediscovered. You probably won't be a whole fossil, but if you're lucky, someone in the future will find enough of your pieces to put you back together again and stick you on a pedestal in the foyer.

Long-term memory

'Unprecedented' is a popular word these days, but nothing is unprecedented if you read some history. The second wave

---

* You might have seen part of this process in action in the limescale in kettles or pipes. Hard water with lots of mineral ions (calcium, magnesium, sulfur, bicarbonate) undergoes a process in which the soluble calcium bicarbonate, when heated, forms insoluble calcium carbonate—the hard, unsightly 'scale'. In a way, you might end up a bit of very old limescale.

of a pandemic, economic crises, revolutions—they all come around regularly enough. Look back further, into geological history, and you'll see the full spectrum of catastrophe: ice ages, heat waves, mass volcanic eruptions, oxygen levels that rise to the point where forests spontaneously combust, then plummet to the point where almost everything dies out;* rocks weathering into soil and soil becoming rock once more. The Earth has seen it all before.

As the world changes, the soil records. Rainfall, gases, temperatures and evidence of what living things were shuffling in or on the soil. Just like archaeologists use objects to reconstruct past civilisations, scientists can use very old soil to glimpse into distant-past ecosystems and climates. To sixteen million years ago, when Australia's inland desert was a woodland. Or ninety million years ago, when Antarctica may have been a rainforest with summer temperatures as warm as nineteen degrees Celsius.†

In Western Australia, high in the Pilbara, lies one of the oldest soils on Earth (now fossilised), formed about 3.5 billion years ago.‡ Back then the world was a drab, rocky place, the air a fog of carbon dioxide and methane. There wasn't anything

---

* There have been five mass extinctions, essentially wiping Earth's hard drive clean, the worst and most famous being the Permian-Triassic that took out ninety-six per cent of marine animals, seventy-five per cent of land dwellers and most insects.
† Soil samples taken from the seabed contained a layer of odd-looking soil that looked more land-like than sea-like. CT scans showed fossilised root systems, so well preserved the scientists could see individual cells and pollen.
‡ The oldest known soil at the moment is in Greenland.

big, except maybe geologically. No trees or animals—it was before land plants. Even if you invented a time-travel machine you couldn't go back without your spacesuit. There was almost no oxygen in the atmosphere, which is why the iron-rich soil is green. When you expose iron to oxygen it rusts, turning a range of reddish hues—like the bright-red banded iron formations in other areas of the Pilbara which are mined for iron ore. This green soil came to be at the beginning of the 'boring billion'—the billion or so years when the atmosphere was largely devoid of oxygen (compared to nearly twenty-one per cent today). There simply wasn't enough oxygen in the atmosphere for the iron to rust.*

This very old soil also holds suggestions of very early life forms. It is more weathered than you would expect from rocks simply bumping into each other, or wetting and drying, so scientists suspect that there was some biological weathering at play. There are clues to this in the fact that the minerals barite and pyrite are present. For barite to form, soil pH needs to fluctuate up and down, dropping to less than 3, about the same pH as stomach acid. Such a low pH doesn't happen without the help of a strong acid such as sulfuric acid, an intensely corrosive substance notorious for its ability to turn a corpse into sludge in about two days.† Pyrite, also known as fool's gold, can react with oxygen to produce sulfuric acid... but if there wasn't oxygen in the atmosphere back then for

---

* Being fossilised into rock has preserved the soil in its pre-oxygen green state.

† British murderer John George Haigh (the Acid Bath Murderer) used sulfuric acid to dissolve at least six of his victims in the 1940s.

iron to rust, there certainly wasn't enough to oxidise pyrite to make sulfuric acid. So where did the acid come from?

Greg Retallack, whom we met briefly at the start of the book, is a world expert on paleosols—very old, sometimes fossilised soils*—and he proposes that it was acid-sulfate weathering caused by sulfur-oxidising bacteria.

Sulfur-oxidising bacteria are single-celled organisms that can chemically oxidise pyrite, using the iron as a fuel source and making sulfuric acid in the process. (These little critters are still around on the Earth, but banished to deep sea vents, cold sulfur lakes and other similarly inhospitable anoxic environments.) Greg's theory is that they worked busily away at the pyrite minerals in the rock, producing sulfuric acid and keeping the pH low enough for barite to form. You don't get this sort of acid-sulfate weathering these days out in the open because there's too much oxygen in the atmosphere.

Old soils tell us what life was like day-to-day during peaceful times, when nothing much was happening except soil was forming, grass was growing and life was living. 'Not like those ambulance-chasing geologists,' says Greg Retallack.

Greg has muddled vowels, a side effect of a life lived across Australia and America: he's spent the last forty years based in Oregon and adventuring across the globe, auger and trowel in hand, studying old soils and reconstructing past climates.

'The soil tells you so much more than rocks alone,' he says. 'What does a lava flow tell you? That tells you a bloody great

---

* The wine-label writers could learn a thing or two from Greg about ancient soils.

lava flow came down and destroyed everything in sight. What does a flood tell you? That there was a flood, and it covered everything in a metre of sediment or so.'

Dramatic geological events are useful to pinpoint changes in Earth's history, but there is more information to be gleaned in the day-to-day—and that is found in the soil. From the soil we can tell whether the landscape was forest or grassland and if it was dry or boggy. Because the soil often preserves some plants and animals we can learn more about the ecosystem as a whole.

Take the example of the Permian-Triassic boundary in the Sydney Basin. It marks the time of the biggest known extinction event, which wiped out the vast majority of life on Earth. We don't know what caused the Great Dying (popular theories involve too much CO_2), but the soil can tell you a lot about what was happening on Earth just before and after. Scattered around the Sydney Basin are a series of old soils that reveal part of the story.

Deep underground, at about twenty metres, lies a thick black layer—coal—laid down in the Permian from about 299 to 252 million years ago. Sydney at the time was a boggy swamp with peat and boreal forest, and it was humid. Fossilised logs trapped in the layer show the trees were about sixteen metres tall; strong growth rings suggest they were seasonally deciduous—something that only happens in places that have distinct seasons. Above the coal is the occasional thick forest soil, with root traces and soil minerals suggesting a cold, temperate environment.

Then the records stop. The dark coal and forest layers cease, as do all fossils. Life, it seems, vanished. The end of the Permian.

Sitting on top of the coal and forest layers are new layers of soil, grey or sometimes a starkly contrasting reddish-orange. This is the Triassic, a new geological era. Head to the beach at Austinmer down near Wollongong and you can see the change in the cliffs. At the bottom of the layer there is still no evidence of life, but clues of widespread deforestation and erosion. Clay breccias* formed by an accumulation of broken rocks and debris suggest a sudden influx of new sediment, washed in by rain and severe soil erosion. What limited flora survived in the early Triassic couldn't stabilise the soil, so the rain eroded the ground. There are coloured mottles and clay skins: mottles form when the soil cycles between wet and dry; clay skins form when clay washes through the soil to form a skin that coats the walls of soil pores. All of this indicates a rainy climate with distinct seasons. Fossils and pollens, and comparison of these soils with similar soils and their associated vegetation today, suggest a cool, temperate broadleaf and conifer forest by 248 million years ago. By sixteen million years ago Sydney was a cool, wet forest with fig trees (*Ficus*) and cabbage tree palms (*Livistona australis*). The soils formed during this time have red nodules, made when the roots of the forest trees drove deep underground, weathering the rock and mobilising nutrients including iron. As Greg Retallack writes

---

* A breccia is a sedimentary rock with lumps of other rocks embedded in it.

in his book *Soil Grown Tall*: 'Dissolved in groundwater, iron came into contact with air and filled tiny pores in the rock and soil with reddish, iron oxide minerals.'

Soils that formed later, beneath the eucalypt woodlands we are more familiar with, are brown and yellow and clayey.

One historical record of particular interest these days is carbon. Carbon dioxide is the common indicator used to track climate changes and communicate findings. Thanks to bubbles of air trapped in Antarctic ice, which act as little time capsules, we have a good picture of CO_2 levels over the past eight hundred thousand years. But going further back in time is much more challenging, so scientists look for different ways to estimate atmospheric CO_2.

The type of carbon in the soil is one clue. Carbon exists in multiple forms, called isotopes. The stable isotopes are carbon-12 (lighter) and carbon-13 (heavier). Carbon-containing material, such as carbonates in the soil, plants or carbon dioxide gas, have different proportions of carbon isotopes. Because atmospheric CO_2 (among other things) influences soil carbonate isotopic composition, analysing soils that were buried and protected gives some clues to historic CO_2 levels. The process is quite complicated. Greg says, 'I haven't worked out a clear way to explain it yet, but very simply, you measure the isotopic composition of carbonates in soils, any organic matter that is there, work out when-abouts in time it came from, then do a whole bunch of math with complicated formulas and you get an atmospheric CO_2 number.'

Carbon isotopes are how we know the current atmospheric CO_2 build-up is human induced. When plants photosynthesise (capturing CO_2 to turn into sugar) they prefer the lighter carbon-12. This means carbon stored in the soil, including the great coal measures on which the modern mining industry is based, mostly contain carbon-12. As CO_2 levels have risen, the ratio of carbon-12 to carbon-13 has dropped (less carbon-13 and more carbon-12). The extra carbon-12 comes from fossil fuels.

Increasing atmospheric CO_2 has a lot of people very worried. Tapping into the soil memory gives us some clues as to how the Earth coped back when CO_2 levels were higher than now.

'Earth had a lot of stretches with four hundred parts per million of CO_2 like now, and it wasn't so bad,' Greg says. 'We're creeping up on the early Pliocene pretty fast here.'

The Pliocene was a time about 2.5 to 5 million years ago when the world was about three degrees warmer than today, and the sun's intensity and CO_2 concentrations were similar to what they are today,* making that era a good analogue for a potential near-future climate. During the Pliocene there weren't any humans, though; the forests and savannahs that became dominant during the Pliocene have little chance of regrowing in this human-centric world.

'We can go up to six hundred parts per million without too much ill effect, like in the Miocene,' says Greg. 'At that point

---

* It's about 420 parts per million (ppm) at the time of writing, and was about 400 ppm in the Pliocene.

CO_2 had a mild fertiliser effect, and tropical plants started to move north and south.'

During the middle Miocene, forests and grasslands covered extensive areas of the planet, partly because more carbon dioxide makes plants grow better: photosynthesis works by using the sun's energy to convert carbon dioxide and water into sugar, their basic building blocks for growth. More CO_2 means more potential building blocks.

But above that level of CO_2 'the rot sets in', as Greg puts it. At 1500 ppm things begin to die. Oxygen starts becoming scarce, particularly oxygen in soil pores, which is critical to keep plants and microbes alive. 'Trees die, swamp plants die and corals die. Then you get mass extinctions, like at the Permian-Triassic boundary.'

During the Permian-Triassic transition, about 252 million years ago, CO_2 went above 2200 ppm.

We're quite a way off 2200 ppm. I hope we don't even get close. But if we do and there's another mass extinction, the soil will record it. And the next round of higher life forms will discover and study it, and perhaps repeat the cycle again.

We were here

Soil remembers the past, even when the landscape has covered it up. Beneath the forests and buildings—even beneath newer soils—lie other soils that remember and record. Sometimes it is just a glimpse into daily life: ash thrown out of a bucket; a walk alongside a drying lake. Other times it records events that also make it into the history books.

What might you deduce if you analysed a sample of forest soil and found quantities of arsenic and zinc ten times higher than expected, alongside elevated phosphorus, chromium, copper, lead, iron, aluminium and mercury? What might have happened here?

You locate yourself on the map. Sztutowo, Poland—on the north coast, not far from white sandy beaches where holiday-makers gather every summer to frolic in the sea. It looks like any other forest: tall, slender trees reaching for the light; leaf litter scattered across the forest floor. Today it might be a nice place for a picnic; eighty years ago, it was a site of horror. The phosphorus comes from bones. Lead from bullets. Mercury from teeth fillings and iron from blood. Trees have sprung up where sixty-five thousand people were buried or incinerated at the Stutthof concentration camp (the current museum and maintained area takes up about twenty per cent of the original site). There is a similar phenomenon at Flanders Field in Belgium, infamous for the mass of red poppies that sprang up after the World War I battles.

Forensic archaeologist Eline Schotsmans—we met her in the Crime chapter—helps find and recover human remains regardless of their age. I spoke to Eline at the University of Wollongong, a couple of hours south of Sydney, where she works at the intersection of archaeology and forensics, special-ising in taphonomy: the influence of the burial environment on the body, and vice versa. An archaeologist by training, Eline ended up in the hybrid field of forensic archaeology when, having noticed that the field was quite rudimentary in continental Europe and especially in her homeland, Belgium, she contacted the police and asked if she could help on cases.

Since her start with the Disaster Victim Investigation team, Eline has ended up working across the globe. She was involved with the search for the remains of Ntare V, the last king of Burundi, assassinated and buried in 1972, although he remains lost—even with the help of a famous DNA specialist, who had identified Louis XIV, and a witness who supposedly took part in the burial. One issue may have been that instead of inter-viewing the witness first to gauge what he remembered, the team took him to the field where the king was thought to be buried and asked him to locate the site. Landscapes can change a lot in twenty-five years, and memories are far from infalli-ble. Because of Burundi's violent history, Eline and team found many human remains—but not those of the king.

In more recent burials there are often clues above ground that something or someone is hidden underground. Clues include vegetation changes such as a dead patch of grass or an unusually healthy patch; a dip in the ground, slumped because the soil was disturbed; or a mound swelling up higher than the surrounding area. Satellites that frequently update aerial imaging can show land surface changes such as crop marks—something Eline looks for every time she takes off or lands in a plane.

Crop marks are usually caused by buried objects influ-encing how plants grow above them. Where the land was excavated for pits or drains or something else, then filled in again with soil over time, there is more soil depth. That means a greater store of water and nutrients, which means plants may grow better than in the surrounding environment. On the other hand, shallower soil above a buried object, such as above wall foundations, can impede root and plant growth. If there is enough of a difference you can sometimes see the outline of a building or channels from above. In a normally rainy climate, crop marks might not be visible until there is drought. Note that soil depth varies naturally: patches of varying grass growth might not be from a human influence at all, though perfectly straight lines are usually a clue to human activity.

In the absence of surface clues there is a range of technology that helps us peer underground. LiDAR (Light Detection and Ranging) measures the shape of the Earth's surface and can reveal subtle indentations on the ground. LiDAR surveys of Angkor Wat and surrounds in Cambodia found multiple cities in the surrounding areas, hidden beneath the forest floor, suggesting a much bigger city and empire than the current Angkor Wat.

As we saw in the Death chapter, soil texture and drainage play a significant role in how well a body is preserved, with decomposition proceeding slower in boggy environments. When Eline studied bones from the wet clay of Flanders Field, the evidence appeared much 'newer' than it was: 'Those soldiers were quite well-preserved,' she says, 'or at least their shoes were...'

Bones are very strong and mainly made from calcium minerals and protein. They are usually the last part of an animal or human to persist in the soil. It would be easy to assume that older bones are more fragile, but age is not necessarily the only factor: the soil plays a role. An example is in the crypt of Bethel Chapel in Sunderland in the UK, last used before 1849. The bodies interred there were buried at approximately the same time, but those on one side of the chapel had well-preserved bones while those on the other side were far more brittle. The difference was partly due to soil pH. The well-preserved bones were in a higher pH soil—about 8.6, alkaline and similar to bones themselves. The brittle bones were in a soil with a pH of about 6.4: on the 'acid' side of the pH scale, with the more acidic soil working to break down the bones faster.

And sometimes the soil conditions work both to destroy the body and to preserve the memory of it. A ghost burial or shadow is an imprint, an area of stained or compacted soil in the shape of what was buried.

In 1939, when local archaeologist Basil Brown excavated mysterious mounds on the property of Edith Pretty at Sutton Hoo in Suffolk, England, he uncovered the ghost of a twenty-seven-metre wooden ship. (The find was recently dramatised in the movie *The Dig*.) The timber did not survive in the strongly acidic (pH 4.5) soil, but it left behind a clear impression pressed into the sand thirteen hundred years later. In the middle of the ship was a burial chamber full of treasures—silverware, gold jewellery and, the most famous find, an ornate iron helmet. Ship burials were rare; whoever buried the ship had to drag it from the water and onto land. The find suggested it was a burial chamber of someone very important, perhaps a king, but there was no body or bones to be found.

The site was quickly taken over by the British Museum and until recently Basil Brown's name was left out of the story. Archaeologists proposed that the ship was a cenotaph, a type of memorial that did not contain human remains. But during further study in the 1960s,* soil analysis beneath the burial chamber found traces of phosphates, suggesting a body had indeed been there but had fully decomposed in the acidic soil. Other mounds at the site held the Sandmen of Sutton Hoo: the death shadows of sixteen people buried all those years ago, and also visible only by the imprint left in the brown sand. We will probably never know exactly who was buried there, just that it was someone who was considered important. More significantly, the find was an important contribution to the recent reassessment of Britain's so-called Dark Ages as a time of art, culture and belief.

As our species expands, we're leaving a bigger and more distinctive imprint in the soil. This stems partly from our skill at digging up elements from one place and putting them in another. Mining phosphorus and potassium for fertiliser to feed an ever-growing population, extracting metals for industry and adornments.

But we've also redistributed some elements that are detrimental when concentrated. Almost all soils contain some

---

* There is also a paper from 1950 published in *Nature* highlighting some 'unusual phosphatic material'.

heavy metals, for example cadmium, chromium, arsenic, lead, zinc, copper or nickel. Heavy metals occur naturally: we use the term 'background levels' to describe the normal amount of metals in a soil before humans started meddling. They are largely determined by what type of rock the soil came from. For example, background lead levels recorded in bushland, in a park and on a farm in New South Wales ranged from 2.7 to 170 milligrams per kilogram.

One sunny afternoon in 2015, Simon (my boss at the time) and I decided to escape the office and spend a pleasant afternoon digging holes. Our mission? To look for chemical evidence of modern life in the soil layers in his garden, located alongside a very busy road—three lanes each way and at least four hours of traffic congestion each day—in a northern suburb of Sydney. For comparison, we collected soil samples from a small patch of remnant native bushland approximately 1.5 kilometres away. The Tumpinyeri bushland is on the same geology and soil as Simon's garden, but is in a quiet, low-traffic area.

Although the garden is set back twenty metres from the road and protected by a stone wall, soil samples collected at a depth of ten centimetres had seven times more zinc (700 mg/kg) and twenty times more lead (460 mg/kg) than the bushland. (This might sound like a lot, but it's not as high as some other urban or industrial levels: 2016 research in an inner suburb of Sydney found average soil lead concentrations of nearly 1000 mg/kg. In the Yunnan province of south-western China, sixty years of lead and zinc smelting deposited just over 8000 mg/kg of zinc, nearly 2500 mg/kg of lead, and 75 mg/kg of cadmium in the surrounding soil.)

As Simon and I dug deeper, the levels of metals dropped. By half a metre down the lead, zinc, nickel and mercury in the garden soil were about the same as the bushland: background levels. Despite the noise and chaos and trucks and 130 years of development, the elemental impact extended only fifty centimetres deep.

This phenomenon, contamination rapidly declining with depth, is quite common, especially if the soil is clayey. Soil texture, organic matter levels and pH affect how well metals and other pollutants can leach deeper underground. Lead, for example, is highly insoluble, meaning it sticks fiercely to soil particles and has a hard time letting go, especially if soil pH is alkaline. I once investigated a site that had lead levels of just over 7000 mg/kg. But because the pH was high, at 8.5, the lead was so tightly bound to the soil that it couldn't wash deeper into the soil or be taken up by plants. The only way that lead was going anywhere was if you dug it up and moved it with the soil. If, however, the soil had started to become more acidic, the lead might have begun to break free. At pH 6.5 there's not too much to worry about; at pH 5.5 that lead is becoming soluble and at greater risk of being more mobile. Organic matter also binds metals, meaning they are less likely to leach deeper into the soil.

Heavy metal accumulation is a well-known problem and scientists are working on ways to undo it. At the moment, one common practice is to excavate contaminated soil and bury it elsewhere, which as you can imagine has its drawbacks. But now there are some interesting biological clean-up methods in development (some earthworms, for example, can chew up cadmium, though it does come at the expense of their health).

One that I find particularly exciting is phytoremediation—using plants and soil microbes to reduce concentrations or toxicity of contaminants—and one particular type of phytoremediation called phytomining is genius.

When researchers noticed the unusual glitter of the eucalyptus leaves in a Western Australian gold-mining area, they discovered that the thirsty tree roots growing up to forty metres deep in the hunt for water were also taking up gold. Could we use plants to suck up metals like mopping up a spill? Quite possibly, as long as we use the right type of plant. There are plants called hyperaccumulators that preferentially take up metals they have no use for. *Berkheya coddii*, for example, a type of thistle with thin yellow petals, likes nickel. If you planted it on a soil with enough nickel to make it worth your while (and of course the right soil chemistry and growing conditions), then recovered the nickel from the plants, you could make over eleven thousand dollars per hectare per harvest, according to one 2009 study from the University of Sydney. *Brassica juncea*, a mustard green used in salads, likes taking up gold. So your mum wasn't lying—greens can be very good for you.

Methods like phytomining are alluring but slow. They can't keep pace with modern life, and as the world gets busier our elemental footprint gets stronger. Though many serious health and environmental hazards are being phased out, modern life will, bit by bit, leave its mark. Old lead pipes and paint flake off and settle in the soil. Chromium spews from combustion—burning gas, coal and oil (in car engines). Zinc comes from galvanised zinc runoff or tyre debris.

We have made such an impact on the Earth that some scientists argue we have started a new geological era—the Anthropocene. Yet despite our influence, in many places the human story in the soil ends about five to ten metres down: a fitting depth for a species that has only existed for the equivalent of a few minutes in the yearly calendar of Earth's history. Beyond that, in the crust that stretches down as far as seventy kilometres, is the rest of Earth's history. Approximately four billion years of tectonic plates colliding, volcanoes erupting, mountains forming and weathering, and life and soil evolving. Within the different layers of soil and rock lies evidence, sometimes obvious, sometimes subtle, of life forms that walked and waddled, respired, transpired and expired.

How long will our era last?

I wonder what visiting life forms will think of our planet, after some sort of disaster, natural or self-induced, wipes out humanity and plants become masters of the Earth once more. When the aliens arrive and point their special analysing ray guns to check for signs of life or hazards, will they consider our soil 'contaminated'? The plants may have covered up the buildings and pavements and farms by then—but if the soil remembers fireplace ash from 150 years ago, it'll certainly remember this busy modern world.

By way of conclusion

WELCOME TO THE end of the book. I hope you've found it to be, as I intended, a light-hearted, intriguing experience that left you feeling excited about soil and all the things it has to offer—outside the realm of agriculture. It is not that agriculture is not important; far from it. Neither you nor I would exist were it not for agriculture. I simply wanted to write something different.

I also wanted to finish the book with a chapter titled 'Future': a speculative, upbeat end note imagining the future of soil. But it turns out that imagining the future means considering the present; and, in a world where nearly half of all habitable land is used for agriculture, it's pretty hard to talk about the present without talking about agriculture. And once I start talking about agriculture there are some unpleasant facts to deal with—about soil degradation, environmental degradation, food security, global warming, economics...

Before I knew it I'd written ten thousand words that made me want to cry when I went back to edit. So I will simply say this: soil has a future in some form. Humanity may not.

'End of the world' narratives are a curious thing. You know the ones: humans have destroyed the world—scorched the sky or salted the earth or just handed things over to a bunch of robots—and in these scenarios humans persist in some post-apocalyptic world, a world of dust, metal and machinery; no greenery in sight.

I do not think this is what it will look like. Humans won't destroy the world. We might damage a lot of it; we might change it significantly; we might even wipe ourselves out. But the Earth? It will recover and find a new way of life. Without meddling humans, the Earth might do quite well. Perhaps plants will become masters of the planet once more. The soil will survive too, in some way or form, keeping a record of these curious two-legged creatures that changed their own habitat so profoundly they only managed to last a few hundred thousand years.

So in the spirit of my original intention, let me propose some modest hopes for whatever future we get to share with our host underfoot.

I would like to see a world where soil exhibits are as commonplace as dinosaur skeletons in natural history museums (museum curators—call me) and soil is part of the broader school curriculum, not just discussed in terms of erosion. Soil science has chemistry, physics and biology—let's make it more common in the classroom.

I want to see more study and understanding to harness the power of soil microbes for the greater good, including new medicines with fewer side effects, fuels and energy sources. More effort put into soil conservation, rehabilitation and appreciation. There are people doing this now, but it is not 'normal'. I want a world where 'soil scientist' isn't such an unfamiliar term that I have to repeat myself or the conversation lurches abruptly to a halt, and where there are more job opportunities for soil scientists, especially outside of agriculture and contaminated land.

I hope there are more books like this, showing how fun and exciting soil can be and focusing on opportunities rather than on doomsday narratives (as justified as some of them might be).

I hope for soil to take its rightful place, recognised alongside plants, water and the atmosphere as critical natural resources. For you, my friend, to go outside and get your hands dirty.

And then perhaps, if we meet at a party and I say I'm a soil scientist, we will have something to talk about.

Acknowledgments

The following people gave their time and expertise to this book. To spice things up, I've listed names in alphabetical order by middle name, but have not included the middle name.

Roy Lawrie, Kelly Dobos, Lynda Williams, Alex McBratney, Bianca Lê, Budiman Minasny, Matt Pepper, Boyd Dent, Daniel Bonn, Robert White, Kevin Hartley, John Davis, Peter Self, Greg Guthrie, Giuseppe Calabrese, Amanda Foxon-Hill, Markus Herderich, Robert Capon, Gwyn Olsen, Greg Retallack, Jarrod Bird, Eline Schotsmans, Lisa Lawrie, Cameron Leckie, Edward Rose, Megan Balks, Rob Fitzpatrick, Wieger Wamelink, Simon Leake, Skye Blackburn.

Thanks to Mum for being my first reader for many of the chapters, especially for trying very hard to research 'paples' (which was a typo of the word apples).

To the team at Text for deciding that a book on soil was something worth publishing. I honestly did not think any publisher would. Glad you proved me wrong.

And to Mike, for your unwavering supporting during the five years it took to get this book out of my head and onto the shelves.

This project is supported by a Writing NSW grant, under a developed funding program run by Writing NSW on behalf of the New South Wales Government and Create NSW.

Notes and references

Life

This chapter draws on interviews with Megan Balks and Greg Retallack.

Two great books on soil formation are *Celebrating Soil* (Springer, 2016) by Megan Balks and Darlene Zabowski and *Soil Grown Tall* by Gregory Retallack (Springer, 2022).

5   (in footnote) that fifty to eighty per cent of atmospheric oxygen comes from phytoplankton in the ocean—National Ocean Service, n.d., *How much oxygen comes from the ocean?* https://oceanservice.noaa.gov/

7   The best description I have found so far is one proposed by Brian Needelman—Needelman, B 2013, 'What are soils?' *Nature Education Knowledge* 4(3):2.

12   these soils could not develop until a bigger life form appeared: grazing animals—Greg Retallack pers. comms; Retallack G J 2013, Global cooling by grassland soils of the geological past and near future. *Annu. Rev. Earth Planet Sci*, vol. 41, pp. 69–86.

16   One example of converting a bare lava flow to a rainforest in Hawai'i goes something like this—the summary is based on the work by Mueller-Dombois D, Boehmer H J 2013, 'Origin of the Hawai'ian rainforest and its transition states in long-term primary succession', *Biogeosciences*, vol. 10, pp. 5171–82.

20   Researchers from the University of Colorado recently collected samples of what appear to be sterile soils from Antarctica—Dragone N B, Diaz M A, Hogg I D, Lyons W B, Jackson W A, Wall D H 2021, 'Exploring the boundaries of microbial habitability in soil', *Journal of Geophysical Research: Biogeosciences*, vol. 126, no. 6.

22   These soil changes mean that the microbes change, with more of the types of microbes you expect in a stomach—Guo Y, Wang N, Li G, Rosas G, Zang J, Ma Y, Liu J, Han W, Cao H 2018, 'Direct and indirect effects of penguin feces on microbiomes in Antarctic ornithogenic soils', *Frontiers in Microbiology*, vol. 9.

Making soil 1: fun and profit

This chapter draws on an interview with Jarod Bird and from experience working with Simon Leake.

The Barangaroo soil design work is summarised in Leake S, Bryce A 2019, 'Design and construction of facsimile yellow kandosols at Barangaroo, Sydney'. In: Vasenev V, Dovletyarova E, Cheng Z, Prokof'eva T, Morel J, Ananyeva N (eds) *Urbanization: Challenge and Opportunity for Soil Functions and Ecosystem Services.* SUITMA 2017. Springer Geography. Springer, Cham.

Although specifications have adapted over time, many elite sports fields and golf course soil was based on the *USGA Recommendations for a Method of Putting Green Construction.*

43 Some literature says 1.6 grams per cubic centimetre—Shipton P and James I 2009, *Guidelines for rolling in cricket,* Cranfield University Centre for Sports Surface Technology, The Grounds Management Association (UK).

44 One set of guidelines from the UK suggest that the more compacted—Shipton P and James I 2009, *Guidelines for rolling in cricket,* Cranfield University Centre for Sports Surface Technology, The Grounds Management Association (UK).

46 he even contemplated suicide after preparing what has been dubbed the 'worst pitch of all time'—Charlie Joseph interview with Paul Weaver for the *Guardian,* 31 Jan 2009.

For more detail on cricket wickets see *Cricket Wickets—Science vs Fiction* by McIntyre and McIntyre, Horticultural Engineering Consultancy, 2001.

The 'bible' I use again and again when working with urban soils is *Growing Media for Ornamental Plants and Turf* by Handreck and Black (UNSW Press, 2010).

Crime

This chapter draws on interviews with Rob Fitzpatrick, Eline Schotsmans and Peter Self.

49 a car boot full of muddy and bloody objects: a shovel, boots, jade bracelet, knife, pine post, towels and bedding—Fitzpatrick R W, Raven M D, Self P G 2017, 'The role of pedology and mineralogy in providing evidence for 5 crime investigations involving a wide range of earth materials', *Episodes Journal of International Geoscience*, vol. 40, no. 2, pp. 148–56.

50 The site operator said no, it wasn't the same type of soil—Porter L, *Written on the Skin*, Pan Macmillan, 2007.

52 just fifteen metres from the large pool of water Rob had pinpointed as the search starting point—Fitzpatrick R W and Raven M D 2012, 'How pedology and mineralogy helped solve a double murder case: using forensics to inspire future generations of soil scientists', *Soil Horizons*, vol. 53, no. 5, pp. 14–29.

54 corpse of Margarethe Filbert was found near Rockenhausen in Bavaria—Chapter 1. A brief history of forensic science and crime scene basics. In Bergslein E. 2012, *An introduction to forensic geoscience*, Blackwell Publishing Ltd.

56 the sand on his shoes compared very strongly with the sand on the beach where Carly was killed—Fitzpatrick R W, Raven M D, Self P G 2017, 'The role of pedology and mineralogy in providing evidence for 5 crime investigations invol.ving a wide range of earth materials', *Episodes Journal of International Geoscience*, vol. 40, no. 2, pp. 148–56.

57 That allowed the prosecution team to show that the alibi that Sinclair came up with in court at the last minute could not possibly have taken place—Lorna Dawson interview with Suzanne Allan for BBC Scotland, 8 October 2016.

59 A train began its journey containing a barrel of silver coins, but when it arrived at the destination the coins were gone and the barrel was full of sand—'Science and Art 1856. Curious use of the microscope'. *Scientific American*, vol. 11, 240.

62 where the 'foetal haemoglobin' (baby's blood) found in the car turned out to be a sound deadener sprayed on when the car was manufactured—Australian Federal Police, 7 May 2018. https://www.afp.gov.au/node/1681

63 despite enjoying the challenge of testing his mushroom identification skills by eating the mushrooms, lived to be nearly eighty—*Clay Minerals Society News*, March 2005.

64 Corryn Rayney was a registrar in the West Australian Supreme Court

who left her home in 2007 to go to a bootscooting class and never returned—Raven M D, Fitzpatrick R W, Self P 2019, 'Trace evidence examination using laboratory and synchrotron X–ray diffraction techniques'. In: Fitzpatrick R W and Donnelly L (eds), *Forensic Soil Science and Geology*. Geological Society, London, Special Publications, 492.

66   conducted enough transference experiments to publish their method in forensic science journals—Murray K, Fitzpatrick R, Bottrill R, Berry R, Kobus H 2016, 'Soil transference patterns on bras: image processing and laboratory dragging experiments', *Forensic Science International*, vol. 258, pp. 88–100.

66   The critical piece of evidence was her pyjama top, which was mysteriously left on the front lawn of a neighbour's house—Fitzpatrick R W, Raven M D 2019, 'The forensic comparison of trace amounts of soil on a pyjama top with hypersulfidic subaqueous soil from a river as evidence in a homicide cold case'. In: Fitzpatrick R W, Donnelly L J (eds) *Forensic Soil Science and Geology*. Geological Society, London, Special Publications, 492.

69   the Australian Guidelines for Conducting Criminal and Environmental Soil Forensics Investigations—Fitzpatrick R W and Raven M D 2016, *Guidelines for Conducting Criminal and Environmental Soil Forensic Investigations (Version 10.1)*. Report, CAFSS_076. Centre for Australian Forensic Soil Science, Adelaide, Australia, p. 46.

70   She picked up nine pig carcasses (as analogues for human corpses)— Schotsmans E M J, Fletcher J N, Denton J, Janaway R C, Wilson A S 2014, 'Long-term effects of hydrated lime and quicklime on the decay of human remains using pig cadavers as human body analogues: field experiments', *Forensic Science International*, vol. 238.

72   former Serbian president Ivan Stambolić was found in a shallow lime-covered grave—*New York Times*, 'Serb gets 40 years in ex-leader's death' 18 July 2005.

72   by the time they were found the lime had formed a death mask— Schotsmans E M J and Van der Voorde W, Chapter 22 'Concealing the crime: the effects of chemicals on human tissues'. In: Schotsmans, E M J, Marquez-Grant N and Forbes S 2017, *Taphonomy of Human*

Remains: Forensic Analysis of the Dead and the Depositional Environment, John Wiley & Sons.

72   **in one case where the Italian mafia executed two victims—** Schotsmans E M J and Van der Voorde W, Chapter 22 'Concealing the crime: the effects of chemicals on human tissues'. *In*: Schotsmans, E M J, Marquez-Grant N and Forbes S 2017, *Taphonomy of Human Remains: Forensic Analysis of the Dead and the Depositional Environment*, John Wiley & Sons.

Evidence from the Earth: Forensic geology and criminal investigation by Raymond Murray (Mountain Press Publishing, 2011) is a wonderful and affordable read. Most of the case information about Craig Smith (pp. 57–8) and the soil substitution–perfume case (pp. 59–60) came from this book.

Wine

This chapter draws on interviews with John Davis, Gwyn Olsen, Robert White and Markus Herderich.

82   **enthusiastic wine journalists say things like 'soils trump climate'** Goode J 2018, *Terroir: When Soils Trump Climate*, WineAnorak.

82   **experiments in which master tasters are served the same wine three times in a row** Derbyshire D 2013, 'Wine tasting: it's junk science', *Guardian* 23 June 2013.

84   **harks back to a time before photosynthesis was discovered—**Maltman A 2013, 'Minerality in wine: a geological perspective', *Journal of Wine Research*, vol. 24, no. 3, pp. 169–81.

86   **When Swiss researchers were investigating why toilets smell so bad—**Starkenmann C, Chappuis C, Niclass Y, Deneulin P 2016, 'Identification of hydrogen disulfanes and hydrogen trisulfanes in H2S bottle, in flint, and in dry mineral white wine', *Journal of Agricultural and Food Chemistry*, vol. 64, pp. 9033–40.

87   **Even some *Penicillium* moulds on corks can synthesise geosmin—** Jung R, Schaefer V 2010, 'Reducing cork taint in wine'. In: Reynolds A G *Managing Wine Quality*, Woodhead Publishing.

89   **too little acid makes a dreary, flat-tasting wine that is prone to spoilage—**Goode J, 2014, *The Science of Wine from Vine to Glass*, University of California Press, Berkeley.

89   Research in Oregon found that as soil pH increased, grape acidity decreased and vice versa—Retallack G J, Burns S F 2016, 'The effects of soil on the taste of wine', *GSA Today*, vol. 26, no. 5.

91   research from Oregon suggests that stressing pinot noir after the onset of ripening makes tastier grapes—Levin A, Jenkins C, Chiginsky J, Williams T, Lake, R 2018, 'Determination of pre- and postveraison water status targets for deficit irrigation of pinot noir in a warm climate', conference poster for American Society of Enology and Viticulture National Conference.

91   A pinot noir can contain over eight hundred organic compounds—Fang Y, Qian M 2005, 'Aroma compounds in Oregon pinot noir wine determined by aroma extract dilution analysis (AEDA)', *Flavour and Fragrance Journal*, vol. 20, pp. 22–9.

92   what gives some Australian shiraz grapes that distinct 'peppery' flavour and how the landscape affects it—Bramley R G V, Siebert T E, Herderich M J, Krstic M P 2017, 'Patterns of within-vineyard spatial variation in the 'pepper' compound rotundone are temporally stable from year to year', *Australian Journal of Grape and Wine Research*, vol. 23, no. 1, pp. 42–7.

95   four major wine-growing regions in California all have different microbiomes—Bokulich N A, Thorngate J H, Richardson P M, Mills D A 2014, 'Microbial biogeography of wine grapes is conditioned by cultivar, vintage, and climate', *PNAS*, vol. 111, no. 1, pp. 139–48.

General wine and soil science books:

•   White R, Krstic M 2019, *Healthy Soils for Healthy Vines*. CSIRO Publishing.

•   White, R 2003, *Soils for Fine Wines*. Oxford University Press.

Health

This chapter draws on interviews with Lynda Williams, Rob Capon and Matt Pepper.

98   the kaolin initiates clotting by triggering a protein called Factor XII—Gegel B, Burgert J, Gasko J, Campbell C, Martens M, Keck J, Reynolds H, Loughren M, Johnson D 2012, 'The effects of QuikClot Combat Gauze and movement on hemorrhage control in a porcine model', *Mil. Med*, vol. 177, pp. 1543–7.

99   Kaolin works about as well as zeolite but without the burning—
     Williams L, Hillier S 2014, 'Kaolins and health: from first grade to
     first aid', *Elements*, vol. 10, pp. 207–11.

99   an exothermic reaction strong enough to cause second degree
     burns—McManus J, Hurtado T, Pusateri A, Knoop K 2007, 'A case
     series describing thermal injury resulting from zeolite use for hemor-
     rhage control in combat operations', *Prehospital emergency care*, vol.
     11, no. 1, pp. 67–71.

102 Staphylococcus and Mycobacterium, while the Argiletz was not.
     Williams L 2017, Geomimicry: harnessing the antibacterial action of
     clays. *Clay Minerals*. 52, 1–24; andWilliams L 2014, Natural anti-
     bacterial clays: historical uses and modern advances, *Clays and Clay
     minerals*, vol. 67, no. 1, pp. 7–24.

107 resulting in more than thirty-five thousand deaths—US Department
     of Health and Human Services Center for Disease Control and
     Prevention (CDC) 2019, Antibiotic resistance threats in the United
     States, Atlanta GA.

108 mid-1980s to the mid-'90s, sixty per cent of newly approved drugs
     came from soil—Brevik E C, Burgess L C 2013, *Soils and Human
     Health*, CRC Press.

108 and are responsible for over half of human antibiotics—Labeda D P,
     Goodfellow M, Brown R et al. 2011, 'Phylogenetic study of the species
     within the family *Streptomycetaceae*', *Antonie van Leeuwenhoek*
     vol. 101, pp. 73–104.

108 *Aspergillus terreus* is a common soil-dwelling fungus—Jahromi
     M F, Liang J B, Ho Y W, Mohamad R, Goh Y M, Shokryazdan
     P 2012, 'Lovastatin production by *Aspergillus terreus* using agro-
     biomass as substrate in solid state fermentation', *J Biomed Biotechnol*
     pp. 1–11.

109 a research trip to the Amazon in 2020 found up to four hundred
     types of fungi—Camila D et al. 2020, 'Advancing biodiversity assess-
     ments with environmental DNA: long-read technologies help reveal
     the drivers of Amazonian fungal diversity', *Ecology and Evol.ution*,
     vol. 10, no. 14, pp. 7509–24.

112 from very pretty locations where the couple wanted to take photos—
     Svarstad H, Dhillion S, Bugge H C 2002, Case Study 6.4. In Laird, S

(ed.) *Biodiversity and Traditional Knowledge: Equitable Partnerships in Practice*. Earthscan.

113 **would cure anything he'd been able to cure while he was alive**—Lidz F 2020, Soil from a Northern Ireland graveyard may lead scientists to a powerful new antibiotic, *Smithsonian Magazine*. Also see Swansea University 2018, 'Bacteria found in ancient Irish soil halts growth of superbugs: new hope for tackling antibiotic resistance', *ScienceDaily*, 27 December 2018.

114 ***Klebsiella pneumoniae* (pneumonia with a mortality rate just below fifty per cent**—Xu L, Sun X, Ma X 2017, 'Systematic review and meta-analysis of mortality of patients infected with carbapenem-resistant Klebsiella pneumoniae', *Ann Clin Microbiol Antimicrob*, vol. 16, no. 1.

116 **the previously unnamed *Eleftheria terrae* which produces teixobactin**—Iyer A, Madder A, Singh I 2019, 'Teixobactins: a new class of 21st century antibiotics to combat multidrug-resistant bacterial pathogens', *Future Microbiology*, vol. 14, no. 6, pp. 457–60.

116 **This method breeds more domesticated microbes that are easier to move into a Petri dish**—Servick K 7 May 2015, 'Microbe found in grassy field contains powerful antibiotic', *Science.org*.

119 **Rook's 'old friends' hypothesis suggests that humans evolved alongside**—Rook G A W, Raison C L, Lowry C A 2014, 'Microbial "old friends", immunoregulation and socioeconomic status', *Clinical and Experimental Immunology*, vol. 177, no. 1, pp. 1–12.

119 **ample evidence from Europe showing that children raised with farm animals**—Kabesch M and Lauener R P 2004, 'Why Old McDonald had a farm but no allergies: genes, environments, and the hygiene hypothesis', *Journal of Leukocyte Biology*, vol. 75, pp. 383–7.

119 **study that compares allergies across the Finnish-Russian border**—Haahtela T, Laatikainen T, Alenius H, Auvinen P, Fyhrquist N, Hanski I, von Hertzen L, Jousilahti P, Kosunen T U, Markelova O, Mäkelä M J, Pantelejev V, Uhanov M, Zilber E, Vartiainen E 2015, Hunt for the origin of allergy—comparing the Finnish and Russian Karelia, *Clinical & Experimental Allergy*, vol. 45, no. 5, pp. 891–901.

120 **one batch of mice on clean bedding; another batch had soil sprinkled on their bedding**—Ottoman N, Ruokolainen L, Suomalainen

A, Sinkko H, Karisola P, Lehtimäki J, Lehto M, Hanski I, Alenius H, Fyhrquist N 2018, 'Soil exposure modifies the gut microbiota and supports immune tolerance in a mouse model', *J Allergy Clin Immunol*, vol. 143, no. 3, pp 1198–1206.

121 **a tray of soil outside a mouse cage and ran a fan**—Liddicoat C, Sydnor H, Cando-Dumancela C, Dresken R, Liu J, Gellie N J C, Mills J G, Young J M, Weyrich L S, Hutchinson M R, Weinstein P, Breed M F 2020, 'Naturally-diverse airborne environmental microbial exposures modulate the gut microbiome and may provide anxiolytic benefits in mice', *Sci Total Environ*.

122 **the Human Microbiome Project**—https://hmpdacc.org/hmp/

122 **domesticated and zoo animals have lower gut microbial diversity than their wild counterparts**—Schultz Marcolla C, Alvarado C S, Willing B P 2019, 'Early life microbial exposure shapes subsequent animal health', *Canadian Journal of Animal Science*, vol. 99, no. 4, pp. 661–77.

Geophagy

124 **evidence from every continent and most cultures at some point in time that people ate soil**—Young S 2011, *Craving Earth: Understanding Pica—the Urge to Eat Clay, Starch, Ice and Chalk*, Columbia University Press.

126 **these tablets could fix everything from stomach ulcers to snakebites**—Spałek K, Spielvogel I 2019, 'The Use of Medicinal Clay from Silesia "Terra sigillata Silesiaca", Central Europe—A New Chance for Natural Medicine?', *Biomed J Sci & Tech Res*, vol. 20, no. 3.

126 **By the 1600s in Paris, Lemnian clay was as popular as mummy flesh**—Huppert G 1999, *The Style of Paris: Renaissance Origins of the French Enlightenment*, Indiana University Press.

126 **When researchers poisoned rats and offered them food or clay or kaolin**—Mitchell D, Wells C, Hoch N, Lind K, Woods S C, Mitchell L K 1976, 'Poison induced pica in rats', *Physiology & Behavior*, vol. 17, no. 4, pp. 691–7.

127 **Peruvian parrots favouring specific soil horizons along the banks of the Manú River in south-eastern Peru**—Gilardi J D, Duffy S S, Munn C A, Tell L A 1999, 'Biochemical functions of geophagy in parrots:

detoxification of dietary toxins and cytoprotective effects', *Journal of Chemical Ecology*, vol. 25. no. 4.

128 **a symptom of a disease called chlorosis, the 'green disease'**— Woywodt A, Kiss A 2002, 'Geophagie: the history of earth-eating', *J R Soc Med*, vol. 95, pp. 143–6.

128 (in footnote) **A recent case of a young girl with a green tinge was treated with iron salt therapy**—Perdahl-Wallace E, Schwartz R H 2006, 'A girl with green complexion an iron deficiency: chlorosis revisited. Case report, March 2006', *Clinical Pediatrics*, vol. 45, no. 2, pp. 187–9.

130 **the clay-rich hills are preferred to the grittier soil on the flats**— Schmidt W E, February 13, 1984, 'Southern practice of eating dirt shows signs of waning', *New York Times* archives.

131 **a British woman lost four as a result of daily soil consumption to ease stress**—'Mum's craving for a cup of dirt a day causes her to lose four teeth', *Metro.co.uk*, 22 June 2021.

Beauty

A good summary paper is Carretero, M 2002, 'Clay minerals and their beneficial effect upon human health. A review', *Applied Clay Science*, vol. 21, pp. 155–63.

137 **can have as many as seventy-five different trace elements adsorbed to their surfaces**—Mbila M 2013, 'Soil minerals, organisms and human health: medicinal uses of soils and soil materials'. In: Brevick & Burgess, *Soils and human health*, CRC Press.

138 **remove excess protein that would otherwise cause a cloudy appearance**—Australian Wine Research Institute, n.d., *Fining Agents*, AWRI, Adelaide, https://www.awri.com.au/industry_support/ winemaking_resources/frequently_asked_questions/fining_agents/

139 **For example, illite is more effective than montmorillonite at adsorbing copper when there is lots of copper around**—Swift R S, McLaren R G 1991, 'Micronutrient adsorption by soils and soil colloids'. In: Bolt G H, de Boodt M F, Hayes M H B, McBride M B (eds), *Interactions at the Soil Colloid–Soil Solution Interface*, Kluwer, Dordrecht.

143 **noticed some kept their feet in warm mud to relieve pain from**

arthritis—R. Kaulitz-Niedeck writes in his book 'Hapsal, ein nordisches al fresko' (1930).

143 benefits of saunas, including on mental and cardiovascular health—Laullanen J A, Laukkanen T, Kunutsor K 2018, 'Cardiovascular and other health benefits of sauna bathing: a review of the evidence', *Mayo Clinic Proceedings*, vol. 93, no. 8, pp. 1111–21.

Making soil 2: extraterrestrial adventures

This chapter draws on interviews with Wieger Wamelink, Giuseppe Calabrese, Skye Blackburn (insect farming) and Bianca Lê (cell meat).

146 or from the briny water on Mars—Rothery D 2020, 'Mars colony: how to make breathable air and fuel from brine—new research', *The Conversation*, 1 December 2020.

150 up to four hundred and fifty kilos of nitrogen per hectare per year if the conditions are right—Pankievicz V C S, Irving T B, Maia L G S et al. 2019, 'Are we there yet? The long walk towards the development of efficient symbiotic associations between nitrogen-fixing bacteria and non-leguminous crops', *BMC Biol*, vol. 77, no. 99.

151 locally grown vegetables have an unfortunate tendency to accumulate higher than desirable—Roba C, Roşu C, Piştea I, Ozunu A, Baciu C 2016, 'Heavy metal content in vegetables and fruits cultivated in Baia Mare mining area (Romania) and health risk assessment', *Environ Sci Pollut Res Int.*, vol. 7, pp. 6062–73.

154 researchers at Colorado State University added the nitrogen-fixing bacteria Sinorhizobium meliloti—Harris F, Dobbs J, Atkins D, Ippolito J A, Stewart J E 2021, 'Soil fertility interactions with *Sinorhizobium*-legume symbiosis in a simulated Martian regolith; effects on nitrogen content and plant health', *PLOS ONE*, vol. 16, no. 9.

156 Experiments on Earth have shown some plants willingly take up perchlorate—Susarla, S, McCutcheon S C 2000, 'Accumulation and fate of perchlorate in plants', Presented at *US Environmental Protection Agency Phytoremediation Conference*, Boston, MA, 1–2 May 2000.

157 tobacco and lettuce plants in particular are known to store perchlorate in their leaves—Sanchez C A, Crump K S, Krieger R I, Khandaker

N R, Gibbs J P 2005, 'Perchlorate and nitrate in leafy vegetables of North America', *Environmental Science and Technology*, vol. 39, no. 24.

157 **A 2013 study suggested mining the perchlorate**—Davila A F, Willson D, Coates J D, McKay C P 2013, 'Perchlorate on Mars: a chemical hazard and a resource for humans', *International Journal of Astrobiology*, vol. 12, no. 4.

157 **Leiden University equipped *E. coli* with a set of genes**—Project by iGEM Team Leiden, https://2016.igem.org/Team:Leiden/Design

161 **multiple studies have found little taste variation between**—Cartier K M S 2018, 'Tests indicate which edible plants could thrive on Mar, *Eos*, *99*, 12 January 2018.

163 **an ingenious system where you use your excrement to feed a 'microbial goo'**—Nelson B 2018, 'How recycled astronaut poop might sustain a mission to Mars', *NBC News*, published online 12 February 2018.

168 **Californian researchers made compressed bricks out of simulated Martian regolith**—St. Fleur N 2017, 'If Mars is colonized, we may not need to ship in the bricks', *New York Times*, published online 28 April 2017.

169 **Many institutions are investigating various agents**—Jakus A, Koube K, Geisendorfer N et al. 2017, 'Robust and elastic lunar and Martian structures from 3D-Printed regolith inks', *Sci Rep,* vol. 7, no. 44931; and Roberts A D, Whittall D R, Breitling R, Takano E, Blaker J J, Hay S, Scrutton N S, 2021, 'Blood, sweat, and tears: extraterrestrial regolith biocomposites with in vivo binders', *Materials Today Bio*, vol. 12.

172 **Penn State developed MarsCrete™ a mixture of basalt rock**—Nazarian S, Duarte J P, Bilén S G, Memari A, Radlinska A, Meisel N, Hojati M 2021, 'Additive Manufacturing of Architectural Structures: An Interplay Between Materials, Systems, and Design'. In: Rodrigues H, Gaspar F, Fernandes P, Mateus A (eds) *Sustainability and Automation in Smart Constructions. Advances in Science, Technology & Innovation*. Springer, Cham.

Death

This chapter draws on interviews with Kevin Hartley and Boyd Dent.

Kevin Hartley's charity, The Earth Funerals Project: www.earthfu-nerals.org/

174 **Australia cremations now account for about sixty-five per cent of end-of-life choices**—Australasian Cemeteries and Crematoria Association, national cremation capacity survey 2020.

174 **In the USA, the burial rate was surpassed by the cremation rate in 2015**—National Funeral Directors Association (NDFA) 2020, 'The future of funerals: COVID-19 restrictions force funeral directors to adapt, propelling the profession forward', *NDFA news release*, published online 7 July 2020.

177 **Sydney's metropolitan gravesites are predicted to be filled up by 2051**—Metropolitan Sydney Cemetery Capacity Report November 2017, *Cemeteries and Crematoria NSW.*

177 **Singapore, the Choa Chu Kang Cemetery Complex is the only**—National Environment Agency Singapore n.d., *Burial, Cremation and Ash Management.* https://www.nea.gov.sg/our-services/after-death/post-death-matters/burial-cremation-and-ash-storage

177–8 **Hong Kong the cremation rate is around ninety per cent, but with thousands of containers of ashes stored in funeral homes**—Teather E K 1999, 'High-rise homes for the ancestors: cremation in Hong Kong', *Geographical Review*, vol. 89, no. 3, pp. 409–30.

178 **there is a process called the 'lift and deepen'**—Dowling J, 'Rest in peace, but not forever', *Age*, 31 December 2012.

178–9 **How soon one might reuse a grave has come up for discussion in reviews**—Report on proceedings before Regulation Committee Inquiry into Cemeteries and Crematoria Amendment Regulation 2018, at Jubilee Room, Parliament House, Sydney, on Friday 21 September 2018.

183 **Vegetarians are more likely to have higher levels of strontium**—Longo UG, Spiezia F, Maffulli N, Denaro V 2008, 'The best athletes in Ancient Rome were vegetarian!', *J Sports Sci Med*, vol. 7, no. 4.

185 **average US cremation creates around the same emissions as two tanks of petrol**—Little B 2019, 'The environmental toll of cremating the dead', *National Geographic*, published online 5 Nov 2019.

185 **Cremation also releases the mercury from anyone who still has**

mercury dental fillings—Piagno H, Afshari R 2020, 'Mercury from crematoriums: human health risk assessment and estimate of total emissions in British Columbia', *Can J Public Health*, vol. 111, no. 6, pp. 1011–19.

188   three hundred natural burial sites in the UK—List of Natural Burial Grounds. http://www.naturaldeath.org.uk/

192   Boyd's PhD thesis, the seminal work on the hydrogeology of cemeteries—Dent B B. The hydrogeological context of cemetery operations and planning in Australia. A thesis submitted to the University of Technology, Sydney for the Degree of Doctor of Philosophy in Science December 2002.

194   The UK has guidelines similar to those Boyd recommends—UK Environment Agency 2020, 'Cemeteries and burials: prevent groundwater pollution', published 14 March 2017, last updated 5 March 2020.

194   As Lee Webster writes in an article for the Green Burial Council—Webster L 2016, 'The science behind green and conventional burial', *Green Burial Council*.

195   the Danish police were called to a peat bog near Silkeborg—Kluger J 2014, 'The bodies in the bogs: an eerie gift from the Iron Age', *Time*, published online 27 July 2014.

197   One review paper from 2015 summarised research—Zychowski J, Bryndal T 2015, 'Impact of cemeteries on groundwater contamination by bacteria and viruses—a review', *Journal of Water and Health IWA Publishing*, vol. 13, no. 2, pp. 285–301.

Also see:

•   Dent B B 2005, 'Vulnerability and the unsaturated zone—the case for cemeteries, where waters meet', *Joint NZ Hydrological Society –AIH (Australia)—NZ Soil Science Society Conference*, Auckland, 28 November–2 December, 2005.

•   Dent B B & Knight M J 1998, 'Cemeteries: a special kind of landfill. The context of their sustainable management', *Groundwater: Sustainable Solutions, Conference of the International Association of Hydrogeologists*, pp. 451–6, Melbourne, February 1998.

War

Thanks to Edward Rose for supplying the papers for (and reviewing) parts of this chapter.

201 Fifteen made it over the sea wall but were then thwarted by the barricades—Lepine M 2019, *D-Day*, Danan Publishing.

202 Napoleon took at least four to help understand the local resources when he invaded Egypt in 1798—Rose E P F, Ehlen J, Lawrence U L 2019, 'Military use of geologists and geology: a historical overview and introduction', *Geological Society, London, Special Publications*, vo. 473, pp. 1–29.

202 'going maps' that showed where tanks could travel across the landscape—Rose T, Clatworthy J, Nathanail P 2006, 'Specialist Maps Prepared by British Military Geologists for the D-Day Landings and Operations in Normandy, 1944', *Cartographic Journal*, vol. 43, no. 2, pp. 117–43.

203 King's appraisal of Normandy geology, 'was, in fact, one of the main factors Inglis J D 1946, 'The work of the Royal Engineers in North-West Europe, 1944–45', *Journal of the Royal United Service Institution*, vol. 91, pp. 176–95. (Reprinted in the *Royal Engineers Journal*, 60, for 1946, 92–112.)

206 reveal a peat layer, and a Roman coin buried in it—Lark M 2008, 'Science on the Normandy Beaches: J. D. Bernal and the Prediction of Soil Trafficability for Operation Overlord', *Soil Survey Horizons*.

206 including one created by the Romans that also showed large areas of peat—Lawson W D 2008, 'Soil Sampling at Sword Beach—Luc-Sur-Mer, France, 1943: how geotechnical engineering influenced the D-Day Invasion and directed the course of modern history', *International Conference on Case Histories in Geotechnical Engineering*, 13.

206 Bernal recalled visiting the beaches on a holiday—Lark M 2008, 'Science on the Normandy Beaches: J. D. Bernal and the prediction of soil trafficability for Operation Overlord', *Soil Survey Horizons*.

207 Sir Malcolm Campbell, who held the land speed record in the 1920s and '30s— Lawson W D 2008, 'Soil Sampling at Sword Beach—Luc-Sur-Mer, France, 1943: how geotechnical engineering influenced

the D-Day Invasion and directed the course of modern history', *International Conference on Case Histories in Geotechnical Engineering*, 13.

208 the equipment list comprised an eighteen-inch soil auger—Trenowden I 1995, *Stealthily by Night: The COPPists Clandestine Beach Reconnaissance and Operations in World War II*, Crecy Books, London.

209 **Strong currents swept Scott-Bowden and Ogden-Smith about 730 metres east of their target—**Lawson W D 2008, 'Soil Sampling at Sword Beach—Luc-Sur-Mer, France, 1943: how geotechnical engineering influenced the D-Day Invasion and directed the course of modern history', *International Conference on Case Histories in Geotechnical Engineering*, 13.

209 **could hear the German revelry coming from the garrison—**Ambrose S E 1994, *D-Day June 6, 1944: The Climactic Battle of World War II*, Simon & Schuster, New York.

209 **(in footnotes) Donald McColville, whose wife didn't find out about his covert soil-sampling activities until the secrecy act ended—**McKenzie S 2012, D-Day's 'forgotten' sand samplers, *BBC News*, 6 June 2012.

210 **gave rise to what was reportedly one of Shotton's favourite reminiscences from the time—**Coope G R 1994, 'Frederick William Shotton: 8 October 1906–21 July 1990', *Biographical Memoirs of Fellows of the Royal Society*, vol. 39, pp. 418–32.

211 **From Bernal's diaries: We jumped out of it onto French soil—**Swann B, Aprahamain F (eds) 1999, *J. D. Bernal: A Life in Science and Politics*, Verso Books.

211 **like flies in amber, were stuck every kind of vehicle except the jeep: tanks, lorries and even DUKWs—**Brown A 2005, *J. D. Bernal: the Sage of Science*, Oxford University Press, Oxford.

212 **there is a photo showing trucks bogged in the sand, captioned in Fred's handwriting—**BBC 2014, WW2 People's War: an archive of World War Two memories—written by the public, gathered by the BBC. 'My Father's D-Day Geologist Role'. Contributed by Anne Black (née Shotton), 22 May 2005.

212 about four per cent of the material was shrapnel and detritus from that fateful day in June 1944—Cooper-van Ingen G 2018, 'War sand by Donald Weber', *Der Grief*, 21 June 2018.

213 with the bedrock and watertable far below the surface—Olson K R, Spiedel D R 2020, 'Review and analysis: successful uses of soil tunnels in medieval and modern warfare and smuggling', *Open Journal of Soil Science*, vol. 10, pp. 194–215.

Stalag Luft III tunnels:

- Crawley A 1956, *Escape from Germany—A History of RAF Escapes During the War*, Collins, London.

- Doyle P, Babits L, Pringle J 2010, Yellow Sands and Penguins: the Soil of 'The Great Escape'. *In:* Landa E, Feller C (eds) *Soil and Culture*, Springer, Dordrecht.

Vietnam tunnels:

- Olson K R and Morton L W 2017, 'Why were the soil tunnels of Cu Chi and Iron Triangle in Vietnam so resilient?', *Open Journal of Soil Science*, vol. 7, pp. 34–51.

Western Front tunnels:

- Doyle P 2012, 'Examples of the influence of groundwater on British military mining in Flanders, 1914–1917', *Geological Society London Special Publications*.

- Goodbody A 2016, 'Tunnelling in the deep', *Mining Magazine*, 13 September 2016.

- Rose E P F 2015, 'Abstract from geology at the Western Front' by T. W. Edgeworth David. *Earth Sciences History*, vol. 34, no. 1, pp. 1–22.

216 When the Germans did a camp inventory after the famous escape— Sky HISTORY. 'The true story of the great escape', https://www. history.co.uk/article/the-true-story-of-the-great-escape

218–19 you would need to stand about 340 metres away from a blast like that to be at a 'safe distance'—thanks to Greg Guthrie for doing this calculation.

220 Captain Martin Greener of the 175th Tunnelling Company recounted

to historian—MacDonald L 1993, *They Called It Passchendaele: The Story of the Battle of Ypres and of the Men Who Fought in It*, Penguin Books.

222 **Australia was able to not only meet our defence needs but help out other nations as well**—Mellor P 1958, Chapter 12 Optical Munitions: *Second World War Official Histories Vol. V—The Role of Science and Industry*. Australian War Memorial Collection.

Quicksand

Thanks to Daniel Bonn for information to help write this chapter.

224 **one in every thirty-five movies (about three per cent) in the 1960s used quicksand**—Engber D 2010, 'Terra Infirma: the rise and fall of quicksand', *Slate Magazine*, 23 August 2010.

226 **A sample of Iranian quicksand**—Khaldoun A, Wegdam G, Eiser E and Bonn D 2006, 'Quicksand!' *Europhysics News*, vol. 37, no. 34, pp. 18–19.

227 **video on YouTube by a man in Northern Arizona showing a patch of quicksand**—Search YouTube for 'Quicksand? Northern Arizona—Little Colorado River!' https://www.youtube.com/watch?v=3HEuyW8TquU

228 **National Geographic has a great video**—Search YouTube for 'Can You Survive Quicksand? I Didn't Know That' https://www.youtube.com/watch?v=a2VJqud3Ls8

229 **Experiments on the Iranian camel-eating quicksand using a model foot**—Khaldoun A, Eiser E, Wegdam G, Bonn D 2005, 'Liquefaction of quicksand under stress', *Nature,* vol. 437, p. 635.

230 **But it has been created in the lab by blowing air through sand**—Lohse D, Rauhé R, Bergmann R, van der Meer D 2004, 'Granular physics: creating a dry variety of quicksand', *Nature*, vol. 432, pp 689–90.

Memory

This chapter draws on interviews with Roy Lawrie, Eline Schotsmans and Greg Retallack.

236 **Ash from *Angophora* species is nearly sixty-three per cent calcium**—Lambert M J 1981, 'Inorganic constituents in wood and bark of NSW

forest tree species', *Forestry Commission of NSW*, research note no. 45.

237 part of Hyde Park was once a dumping ground for rubbish—Roy Lawrie pers. comm., 2021.

239 scientist Karen Chin concluded had been excreted by a T-rex—Chin K, Tokaryk T, Erickson G, Calk L 1998, 'A king-sized theropod coprolite', *Nature,* vol. 393, pp. 680–2.

239–40 protozoan cysts and nematode eggs indicating the presence of intestinal parasites—Poinar G, Boucot A J 2006, 'Evidence of intestinal parasites of dinosaurs', *Parisitology*, vol. 133, pp. 245–9.

241 The age of the Mungo prints was established by measuring the amount of radiation stored in the soil—Webb S, Cupper M L, Robins R 2006, 'Pleistocene human footprints from the Willandra Lakes, southeastern Australia', *Journal of Human Evol.ution*, vol. 50.

241 hadn't seen light since they were buried by sand blown in about twenty thousand years ago—Westaway M C, Cupper M L, Johnston H, Graham I 2013, 'The Willandra Fossil Trackway: assessment of ground penetrating radar survey results and additional OSL dating at a unique Australian site', *Australian Archaeology*, no. 76, pp. 84–9.

241–2   one hunter may have run at thirty-seven kilometres per hour—Webb S 2007, 'Further research of the Willandra Lakes fossil footprint site, southeastern Australia', *Journal of Human Evol.ution*, vol. 52, no. 4, pp. 711–15.

242 One child moved in the way children do: meandering in one direction then suddenly another—Westaway M C, Cupper M L, Johnston H, Graham I 2013, 'The Willandra Fossil Trackway: assessment of ground penetrating radar survey results and additional OSL dating at a unique Australian site', *Australian Archaeology*, no. 76, pp. 84–9.

To read more in general about the Mungo Footprints see: http://www.visitmungo.com.au/footprint-makers#who-made-footprints

246 when Antarctica may have been a rainforest—Klages J P, Salzmann U, Bickert T et al. 2020, 'Temperate rainforests near the South Pole during peak Cretaceous warmth', *Nature*, vol. 580, pp. 81–6.

247 that the minerals barite and pyrite are present—Retallack G J 2018, 'The oldest known paleosol profiles on Earth: 3.46Ga Panorama

Formation, Western Australia', *Palaeogeography, Palaeoclimatology, Palaeoecology,* vol. 489, pp. 230–48.

249 **Fossilised logs trapped in the layer show the trees were about sixteen metres tall**—Retallack G J 1999, 'Post-apocalyptic greenhouse paleoclimate revealed by earliest Triassic paleosols in the Sydney Basin, Australia', *Bulletin of the Geological Society of America,* vol. 111, no.1, pp. 52–70.

250 **still no evidence of life, but clues of widespread deforestation and erosion**—Retallack G J 1999, 'Post-apocalyptic greenhouse paleoclimate revealed by earliest Triassic paleosols in the Sydney Basin, Australia', *Bulletin of the Geological Society of America,* vol. 111, no.1, pp. 52–70.

250 **suggest a cool, temperate broadleaf and conifer forest by 248 million years ago**—Retallack G 2022, *Soil Grown Tall,* Springer.

253 **grasslands covered extensive areas of the planet, partly because more carbon dioxide**—Kürschner W M, Kvaček Z, Dilcher D L 2008, 'The impact of Miocene atmospheric carbon dioxide fluctuations on climate and the evol.ution of terrestrial ecosystems', *PNAS,* vol. 105, no. 2, pp 449–53.

253 **During the Permian-Triassic transition, about 252 million years ago, CO2 went above 2200 ppm**—Wu Y, Chu D, Tong J et al. 2021, 'Sixfold increase of atmospheric CO2 during the Permian–Triassic mass extinction', *Nat Communications,* vol. 12, no. 2137.

254 **a sample of forest soil and found quantities of arsenic and zinc ten times higher than expected**—Charzynski P, Markiewicz M, Majorek M, Bednarek R 2015, 'Geochemical assessment of soils in the German Nazi concentration camp in Stutthof (Northern Poland)', *Soil Science and Plant Nutrition.* vol. 61, pp. 47–54.

256 **LiDAR surveys of Angkor Wat and surrounds in Cambodia**—Wallace J 2016, 'Laser scans unveil a network of ancient cities in Cambodia', *New York Times,* 19 September 2016.

257 **The timber did not survive in the strongly acidic (pH 4.5) soil**—Barker H 1950, 'Unusual phosphatic material in the Sutton Hoo ship burial', *Nature,* vol. 166, no. 348.

258 **during further study in the 1960s**—UK National Trust n.d., History

of Archaeology at Sutton Hoo, https://www.nationaltrust.org.uk/sutton-hoo/features/history-of-archaeology-at-sutton-hoo

258 (in footnote) **There is also a paper from 1950 published in *Nature* highlighting some 'unusual phosphatic material'**—Barker H 1950, 'Unusual phosphatic material in the Sutton Hoo ship burial', *Nature*, vol. 166, no. 348.

259 **background lead levels recorded in bushland, in a park and on a farm in New South Wales**—Olszowy P, Torr P, Imray P 1995, 'Trace element concentrations in soils from rural and urban areas of Australia', *Contaminated Sites Monograph Series*, no. 4.

259 **2016 research in an inner suburb of Sydney found average soil lead concentrations of nearly 1000 mg/kg**—Rouillon M, Kristensen L, Taylor M P, Harvey P, George S G 2017, 'Elevated lead levels in Sydney back yards: here's what you can do', *The Conversation*, 17 January 2017.

259 **In the Yunnan province of south-western China, sixty years of lead and zinc smelting**—Li P, Lin C, Cheng H, Duan X, Lei K 2015, 'Contamination and health risks of soil heavy metals around a lead/zinc smelter in southwestern China', *Ecotoxicology and Environmental Safety*, vol. 113, pp. 391–9.

261 **some earthworms, for example, can chew up cadmium, though it does come at the expense of their health**—Wu Y, Chen C, Wang G, Xiong B, Zhou W, Xue F, Qi W, Qiu C, Liu Z 2020, 'Mechanism underlying earthworm on the remediation of cadmium-contaminated soil', *The Science of the Total Environment, vol. 728*.

261 **the unusual glitter of the eucalyptus leaves in a Western Australian gold-mining area**—Lintern M, Anand R, Ryan C et al. 2013, 'Natural gold particles in Eucalyptus leaves and their relevance to exploration for buried gold deposits', *Nat Commun*, vol. 4.

261 **make over eleven thousand dollars per hectare per harvest, according to one 2009 study**—Harris A T, Naidoo K, Nokes J, Walker T, Orton F 2009, 'Indicative assessment of the feasibility of Ni and Au phytomining in Australia', *Journal of Cleaner Production*, vol. 17, no. 2, pp. 194–200.

Index